Rick

I Thought This was one of
The BEST HISTORIES of CYBER-
I'VE READ. —

Maybe you will ENJOY it ALSO. —

Love
DAD

P.S. MAYBE CAUSE I CAN
RELATE to A LOT OF THE STORY. !

THE SECRET HISTORY OF CYBER WAR

DARK TERRITORY

FRED KAPLAN

SIMON & SCHUSTER

NEW YORK LONDON TORONTO SYDNEY NEW DELHI

Simon & Schuster
1230 Avenue of the Americas
New York, NY 10020

Copyright © 2016 by Fred Kaplan

First Simon & Schuster hardcover edition March 2016

SIMON & SCHUSTER and colophon are registered trademarks
of Simon & Schuster, Inc.

For information about special discounts for bulk purchases,
please contact Simon & Schuster Special Sales at 1-866-506-1949 or
business@simonandschuster.com.

The Simon & Schuster Speakers Bureau can bring authors to your live event.
For more information or to book an event, contact the
Simon & Schuster Speakers Bureau at 1-866-248-3049 or
visit our website at www.simonspeakers.com.

Book design by Ellen R. Sasahara

Manufactured in the United States of America

1 3 5 7 9 10 8 6 4 2

Library of Congress Cataloging-in-Publication Data
Kaplan, Fred M.
Dark territory : the secret history of cyber war / Fred Kaplan.
New York, NY : Simon & Schuster, 2016. | Includes
bibliographical references and index.
LCSH: Cyberterrorism—Prevention—United States—History. | BISAC:
TECHNOLOGY & ENGINEERING / Military Science. | COMPUTERS /
Security /
Viruses. | HISTORY / Military / General.
Classification: LCC HV6773.15.C97 K37 2016 | DDC 363.325—dc23 LC record
available at http://lccn.loc.gov/2015027335

ISBN 978-1-4767-6325-5
ISBN 978-1-4767-6327-9 (ebook)

for Brooke Gladstone

CONTENTS

DARK TERRITORY

CHAPTER 1

"COULD SOMETHING LIKE THIS REALLY HAPPEN?"

I T was Saturday, June 4, 1983, and President Ronald Reagan spent the day at Camp David, relaxing, reading some papers, then, after dinner, settling in, as he often did, to watch a movie. That night's feature was *WarGames*, starring Matthew Broderick as a tech-whiz teenager who unwittingly hacks into the main computer at NORAD, the North American Aerospace Defense Command, and, thinking that he's playing a new computer game, nearly triggers World War III.

The following Wednesday morning, back in the White House, Reagan met with the secretaries of state, defense, and treasury, his national security staff, the chairman of the Joint Chiefs of Staff, and sixteen prominent members of Congress, to discuss a new type of nuclear missile and the prospect of arms talks with the Russians. But he couldn't get that movie out of his mind. At one point, he put down his index cards and asked if anyone else had seen it. Nobody had (it had just opened in theaters the previous Friday), so he launched into a detailed summary of its plot. Some of the legislators looked around the room with suppressed smiles or arched eyebrows. Not quite three months earlier, Reagan had delivered his "Star Wars" speech, calling on scientists to develop laser weapons that, in the event of war, could shoot down Soviet nuclear missiles as they darted toward America. The idea was widely dismissed as nutty. What was the old man up to now?

After finishing his synopsis, Reagan turned to General John Vessey, the chairman of the Joint Chiefs, the U.S. military's top officer, and asked, "Could something like this really happen?" Could someone break into our most sensitive computers?

Vessey, who'd grown accustomed to such queries, said he would look into it.

One week later, the general came back to the White House with his answer. *WarGames*, it turned out, wasn't at all far-fetched. "Mr. President," he said, "the problem is much worse than you think."

Reagan's question set off a string of interagency memos, working groups, studies, and meetings, which culminated, fifteen months later, in a confidential national security decision directive, NSDD-145, signed September 17, 1984, titled "National Policy on Telecommunications and Automated Information Systems Security."

It was a prescient document. The first laptop computers had barely hit the market, the first public Internet providers wouldn't come online for another few years. Yet the authors of NSDD-145 noted that these new devices—which government agencies and high-tech industries had started buying at a rapid clip—were "highly susceptible to interception, unauthorized electronic access, and related forms of technical exploitation." Hostile foreign intelligence agencies were "extensively" hacking into these services already, and "terrorist groups and criminal elements" had the ability to do so as well.

This sequence of events—Reagan's oddball question to General Vessey, followed by a pathbreaking policy document—marked the first time that an American president, or a White House directive, discussed what would come to be called "cyber warfare."

The commotion, for now, was short-lived. NSDD-145 placed the National Security Agency in charge of securing all computer servers and networks in the United States, and, for many, that went too far. The NSA was America's largest and most secretive intelligence agency. (Insiders joked that the initials stood for "No Such Agency.") Established in 1952 to intercept foreign communications, it was expressly forbidden from spying on Americans. Civil liberties advocates in Congress were not about to let a presidential decree blur this distinction.

And so the issue vanished, at least in the realm of high-level politics. When it reemerged a dozen years later, after a spate of actual cyber intrusions during Bill Clinton's presidency, enough time had passed that the senior officials of the day—who didn't remember, if they'd ever known of, NSDD-145—were shocked by the nation's seemingly sudden vulnerability to this seemingly brand-new threat.

When the White House again changed hands (and political parties) with the election of George W. Bush, the issue receded once more, at least to the public eye, especially after the terrorist attacks of September 11, 2001, which killed three thousand Americans. Few cared about hypothetical cyber wars when the nation was charging into real ones with bullets and bombs.

But behind closed doors, the Bush administration was weaving cyber war techniques with conventional war plans, and so were the military establishments of several other nations, friendly and otherwise, as the Internet spread to the globe's far-flung corners. Cyber war emerged as a mutual threat *and* opportunity, a tool of espionage and a weapon of war, that foes could use to hurt America and that America could use to hurt its foes.

During Barack Obama's presidency, cyber warfare took off, emerging as one of the few sectors of the defense budget that soared while others stayed stagnant or declined. In 2009, Obama's first secretary of defense, Robert Gates, a holdover from the Bush years, created a dedicated Cyber Command. In its first three years, the command's annual budget tripled, from $2.7 billion to $7 billion (plus another $7 billion for cyber activities in the military services, all told), while the ranks of its cyber attack teams swelled from 900 personnel to 4,000, with 14,000 foreseen by the end of the decade.

The cyber field swelled worldwide. By the midpoint of Obama's presidency, more than twenty nations had formed cyber warfare units in their militaries. Each day brought new reports of cyber attacks, mounted by China, Russia, Iran, Syria, North Korea, and others, against the computer networks of not just the Pentagon and defense contractors but also banks, retailers, factories, electric power grids, waterworks—everything connected to a computer network, and, by the early twenty-first century, that included nearly everything. And, though much less publicized, the United States and a few other Western powers were mounting cyber attacks on other nations' computer networks, too.

In one sense, these intrusions were nothing new. As far back as Roman times, armies intercepted enemy communications. In the American Civil War, Union and Confederate generals used the new telegraph machines to send false orders to the enemy. During World War II, British and American cryptographers broke German and Japanese codes, a crucial ingredient (kept secret for many years after) in the Allied victory. In the first few decades of the Cold War, American and Russian spies routinely intercepted each other's radio signals, microwave transmissions, and telephone calls, not just to gather intelligence about intentions and capabilities but, still more, to gain an advantage in the titanic war to come.

In other ways, though, information warfare took on a whole new dimension in the cyber age. Until the new era, the crews gathering SIGINT—signals intelligence—tapped phone lines and swept the skies for stray electrons, but that's all they could do: *listen* to conversations, *retrieve* the signals. In the cyber age, once they hacked a computer, they could prowl the entire network connected to it; and, once inside the network, they could not only read or download scads of information; they could change its content—disrupt, corrupt, or erase it—and mislead or disorient the officials who relied on it.

Once the workings of almost everything in life were controlled by or through computers—the guidance systems of smart bombs, the centrifuges in a uranium-enrichment lab, the control valves of a dam, the financial transactions of banks, even the internal mechanics of cars, thermostats, burglary alarms, toasters—hacking into a network gave a spy or cyber warrior the power to control those centrifuges, dams, and transactions: to switch their settings, slow them down, speed them up, or disable, even destroy them.

This damage was wreaked remotely; the attackers might be half a world away from the target. And unlike the atomic bomb or the intercontinental ballistic missile, which had long ago erased the immunity of distance, a cyber weapon didn't require a large-scale industrial project or a campus of brilliant scientists; all it took to build one was a roomful of computers and a small corps of people trained to use them.

There was another shift: the World Wide Web, as it came to be called, was just that—a network stretched across the globe. Many classified programs ran on this same network; the difference was that their contents were encrypted, but this only meant that, with enough time and effort, they could be decrypted or otherwise penetrated, too. In the old days, if spies wanted to tap a phone, they put a device on a single circuit. In the cyber era, Internet traffic moved at lightning speed, in digital packets, often interspersed with packets containing other people's traffic, so a terrorist's emails or cell phone chatter couldn't be extracted so delicately; everyone's chatter and traffic got tossed in the dragnet, placed, potentially, under the ever-watchful eye.

The expectation arose that wars of the future were bound to be, at least in part, cyber wars; cyberspace was officially labeled a "domain" of warfare, like air, land, sea, and outer space. And because of the seamless worldwide network, the packets, and the Internet of Things, cyber war would involve not just soldiers, sailors, and pilots but, inexorably, the rest of us. When cyberspace is everywhere, cyber war can seep through every digital pore.

During the transitions between presidents, the ideas of cyber warfare were dismissed, ignored, or forgotten, but they never disappeared. All along, and even before Ronald Reagan watched *War-Games*, esoteric enclaves of the national-security bureaucracy toiled away on fixing—and, still more, exploiting—the flaws in computer software.

General Jack Vessey could answer Reagan's question so quickly—within a week of the meeting on June 8, 1983, where the president asked if someone could really hack the military's computers, like the kid in that movie—because he took the question to a man named Donald Latham. Latham was the assistant secretary of defense for command, control, communications, and intelligence—ASD(C3I), for short—and, as such, the Pentagon's liaison with the National Security Agency, which itself was an extremely secret part of the Department of Defense. Spread out among a vast complex of shuttered buildings in Fort Meade, Maryland, surrounded by armed guards and high gates, the NSA was much larger, better funded, and more densely populated than the more famous Central Intelligence Agency in Langley, Virginia. Like many past (and future) officials in his position, Latham had once worked at the NSA, still had contacts there, and knew the ins and outs of signals intelligence and how to break into communications systems here and abroad.

There were also top secret communications-intelligence bureaus of the individual armed services: the Air Intelligence Agency (later called the Air Force Information Warfare Center) at Kelly Air Force Base in San Antonio, Texas; the 609th Information Warfare Squadron at Shaw Air Force Base in Sumter, South Carolina; scattered cryptology labs in the Navy; the CIA's Critical Defense Technologies Division; the Special Technological Operations Division of J-39, a little known office in the Pentagon's Joint Staff (entry required dialing the combination locks on two metal doors). They all fed to and from the same centers of beyond-top-secret wizardry, some of it homegrown, some manufactured by ESL, Inc. and other specialized private contractors. And they all interacted, in one way or another, with the NSA.

When Reagan asked Vessey if someone could really hack into the military's computers, it was far from the first time the question had been asked. To those who would write NSDD-145, the question was already very old, as old as the Internet itself.

———

In the late 1960s, long before Ronald Reagan watched *WarGames*, the Defense Department undertook a program called the ARPANET. Its direct sponsor, ARPA (which stood for Advanced Research Projects Agency), was in charge of developing futuristic weapons for the U.S. military. The idea behind ARPANET was to let the agency's contractors—scientists at labs and universities across the country—share data, papers, and discoveries on the same network. Since more and more researchers were using computers, the idea made sense. As things stood, the director of ARPA had to have as many computer consoles in his office as there were contractors out in the field, each hooked up to a separate telephone modem—one to communicate with UCLA, another with the Stanford Research Institute, another with the University of Utah, and so forth. A single network, linking them all, would not only be more economical, it would also let scientists around the country exchange data more freely and openly; it would be a boon to scientific research.

In April 1967, shortly before ARPANET's rollout, an engineer named Willis Ware wrote a paper called "Security and Privacy in Computer Systems" and delivered it at the semiannual Joint Computer Conference in New York City. Ware was a pioneer in the field of computers, dating back to the late 1940s, when there barely was such a field. At Princeton's Institute for Advanced Studies, he'd been a protégé of John von Neumann, helping design one of the first electrical computers. For years now, he headed the computer science department at the RAND Corporation, an Air Force–funded think tank in Santa Monica, California. He well understood the point of ARPANET, lauded its goals, admired its ambition; but he was worried about some implications that its managers had overlooked.

In his paper, Ware laid out the risks of what he called "resource-sharing" and "on-line" computer networks. As long as computers stood in isolated chambers, security wouldn't be a problem. But once multiple users could access data from unprotected locations, anyone with certain skills could hack into the network—and after hacking into one part of the network, he could roam at will.

Ware was particularly concerned about this problem because he knew that defense contractors had been asking the Pentagon for permission to store classified and unclassified files on a single computer. Again, on one level, the idea made sense: computers were expensive; commingling all the data would save lots of money. But in the impending age of ARPANET, this practice could prove disastrous. A spy who hacked into unclassified networks, which were entirely unprotected, could find "back doors" leading to the classified sections. In other words, the very existence of a network created sensitive vulnerabilities; it would no longer be possible to keep secrets.

Stephen Lukasik, ARPA's deputy director and the supervisor of the ARPANET program, took the paper to Lawrence Roberts, the project's chief scientist. Two years earlier, Roberts had designed a communications link, over a 1200-baud phone line, between a computer at MIT's Lincoln Lab, where he was working at the time, and a colleague's computer in Santa Monica. It was the first time anyone had pulled off the feat: he was, in effect, the Alexander Graham Bell of the computer age. Yet Roberts hadn't thought about the security of this hookup. In fact, Ware's paper annoyed him. He begged Lukasik not to saddle his team with a security requirement: it would be like telling the Wright brothers that their first airplane at Kitty Hawk had to fly fifty miles while carrying twenty passengers. Let's do this step by step, Roberts said. It had been hard enough to get the system to *work*; the Russians wouldn't be able to build something like this for decades.

He was right; it *would* take the Russians (and the Chinese and others) decades—about three decades—to develop their versions of the ARPANET and the technology to hack into America's. Meanwhile, vast systems and networks would sprout up throughout the United States and much of the world, without any provisions for security.

Over the next forty years, Ware would serve as a consultant on government boards and commissions dealing with computer security and privacy. In 1980, Lawrence Lasker and Walter Parkes, former Yale classmates in their late twenties, were writing the screenplay for the film that would come to be called *WarGames*. They were uncertain about some of the plotline's plausibility. A hacker friend had told them about "demon-dialing" (also called "war-dialing"), in which a telephone modem searched for other nearby modems by automatically dialing each phone number in a local area code and letting it ring twice before moving on to the next number. If a modem answered, it would squawk; the demon-dialing software would record that number, and the hacker would call it back later. (This was the way that early computer geeks found one another: a pre-Internet form of web trolling.) In the screenplay, this was how their whiz-kid hero breaks into the NORAD computer. But Lasker and Parkes wondered whether this was possible: wouldn't a military computer be closed off to public phone lines?

Lasker lived in Santa Monica, a few blocks from RAND. Figuring that someone there might be helpful, he called the public affairs officer, who put him in touch with Ware, who invited the pair to his office.

They'd found the right man. Not only had Ware long known about the myriad vulnerabilities of computer networks, he'd helped design the software program at NORAD. And for someone so steeped in the world of big secrets, Ware was remarkably *open*, even friendly. He looked like Jiminy Cricket from the Disney cartoon film of *Pinocchio*, and he acted a bit like him, too: excitable, quick-witted, quick to laugh.

Listening to the pair's questions, Ware waved off their worries. Yes, he told them, the NORAD computer was supposed to be closed, but some officers wanted to work from home on the weekend, so they'd leave a port open. Anyone could get in, if the right number was dialed. Ware was letting the fledgling screenwriters in on a secret that few of his colleagues knew. The only computer that's completely secure, he told them with a mischievous smile, is a computer that no one can use.

Ware gave Lasker and Parkes the confidence to move forward with their project. They weren't interested in writing sheer fantasy; they wanted to imbue even the unlikeliest of plot twists with a grain of authenticity, and Ware gave them that. It was fitting that the scenario of *WarGames*, which aroused Ronald Reagan's curiosity and led to the first national policy on reducing the vulnerability of computers, was in good part the creation of the man who'd first warned that they were vulnerable.

Ware couldn't say so, but besides working for RAND, he also served on the Scientific Advisory Board of the National Security Agency. He knew the many ways in which the NSA's signals intelligence crews were piercing the shields—penetrating the radio and telephone communications—of the Russian and Chinese military establishments. Neither of those countries had computers at the time, but ARPANET was wired through dial-up modems—through phone lines. Ware knew that Russia or China could hack into America's phone lines, and thus into ARPANET, with the same bag of tricks that America was using to hack into their phone lines.

In other words, what the United States was doing to its enemies, its enemies could also do to the United States—maybe not right now, but someday soon.

———

The National Security Agency had its roots in the First World War. In August 1917, shortly after joining the fight, the United States government created Military Intelligence Branch 8, or MI-8, devoted to deciphering German telegraph signals. The unit stayed open even after the war, under the dual auspices of the war and state departments, inside an inconspicuous building in New York City that its denizens called the Black Chamber. The unit, whose cover name was the Code Compilation Company, monitored communications of suspected subversives; its biggest coup was persuading Western Union to provide access to all the telegrams coming over its wires. The Black Chamber was finally shut down in 1929, after Secretary of State Henry Stimson proclaimed, "Gentlemen don't read each other's mail." But the practice was revived, with the outbreak of World War II, as the Signal Security Agency, which, along with British counterparts, broke the codes of German and Japanese communications—a feat that helped the Allies win the war. Afterward, it morphed into the Army Security Agency, then the multiservice Armed Forces Security Agency, then in 1952—when President Harry Truman realized the services weren't cooperating with one another—a unified code-breaking organization called the National Security Agency.

Throughout the Cold War, the NSA set up bases around the world—huge antennas, dishes, and listening stations in the United Kingdom, Canada, Japan, Germany, Australia, and New Zealand—to intercept, translate, and analyze all manner of communications inside the Soviet Union. The CIA and the Air Force flew electronic-intelligence airplanes along, and sometimes across, the Soviet border, picking up signals as well. In still riskier operations, the Navy sent submarines, equipped with antennas and cables, into Soviet harbors.

In the early years of the Cold War, they were all listening mainly to radio signals, which bounced off the ionosphere all around the globe; a powerful antenna or large dish could pick up signals from just about anyplace. Then, in the 1970s, the Russians started switching to microwave transmissions, which beamed across much shorter distances; receivers had to be in the beam's line of sight to intercept it. So the NSA created joint programs, sending spies from the CIA or other agencies across enemy lines, mainly in the Warsaw Pact nations of Eastern Europe, to erect listening posts that looked like highway markers, telephone poles, or other mundane objects.

Inside Moscow, on the tenth floor of the American embassy, the NSA installed a vast array of electronic intelligence gear. In a city of few skyscrapers, the tenth floor offered a panoramic view. Microwave receivers scooped up phone conversations between top Soviet officials—including Chairman Leonid Brezhnev himself—as they rode around the city in their limousines.

The KGB suspected something peculiar was going on up there. On January 20, 1978, Bobby Ray Inman, the NSA director, was awakened by a phone call from Warren Christopher, the deputy secretary of state. A fire had erupted in the Moscow embassy, and the local fire chief was saying he wouldn't put it out unless he was given access to the tenth floor. Christopher asked Inman what he should do.

Inman replied, "Let it burn." (The firefighters eventually put it out anyway. It was one of several fires that mysteriously broke out in the embassy during that era.)

By 1980, the last full year of Jimmy Carter's presidency, the American spy agencies had penetrated the Soviet military machine so deeply, from so many angles, that analysts were able to piece together a near-complete picture of its operations, patterns, strengths, and weaknesses. And they realized that, despite its enormous buildup in troops and tanks and missiles, the Soviet military was extremely vulnerable.

The fatal gaps lay in the communications links of its command-control systems—the means by which radar operators tracked incoming planes and missiles, general officers sent out orders, and Kremlin higher-ups decided whether to go to war. And once American SIGINT crews were inside Soviet command-control, they could not only learn what the Russians were up to, which was valuable enough; they could also insert false information, disrupt the command signals, even shut them off. These disruptions might not win a war by themselves, but they could tip the balance, sowing confusion among Soviet officers, making them distrust the intelligence they were seeing and the orders they were receiving—which, in the best of scenarios, might stop them from launching a war in the first place.

The Russians, by now, had learned to encrypt their most vital command-control channels, but the NSA figured out how to break the codes, at least some of them. When cryptologists of whatever nationality coded a signal, they usually made a mistake here and there, leaving some passages in plain text. One way to break the code was to find the mistake, work backward to see how that passage—say, an often-used greeting or routine military jargon—had been encrypted in previous communiqués, then unravel the code from there.

Bobby Ray Inman had been director of naval intelligence before he took over the NSA in 1977, at the start of President Carter's term. Even back then, he and his aides had fiddled with encryption puzzles. Now with the NSA's vast secret budget at his disposal, Inman went at the task with full steam. In order to compare encrypted passages with mistakes in the clear, he needed machines that could store a lot of data and process it at high speed. For many years, the NSA had been building computers—vast corridors were filled with them—but this new task exceeded their capacity. So, early on in his term as director, Inman started a program called the Bauded Signals Upgrade, which involved the first "supercomputer." The machine cost more than a billion dollars, and its usefulness was short-lived: once the Soviets caught on that their codes had been broken, they would devise new ones, and the NSA code breakers would have to start over. But for a brief period of Russian obliviousness, the BSU helped break enough high-level codes that, combined with knowledge gained from other penetrations, the United States acquired an edge—potentially a decisive edge—in the deadliest dimension of the Cold War competition.

Inman had a strong ally in the Pentagon's top scientist, William Perry. For a quarter century, Perry had immersed himself in precisely this way of thinking. After his Army service at the end of World War II, Perry earned advanced degrees in mathematics and took a job at Sylvania Labs, one of the many high-tech defense contractors sprouting up in Northern California, the area that would later be called Silicon Valley. While many of these firms were designing radar and weapons systems, Sylvania specialized in electronic *counter*measures—devices that jammed, diffracted, or disabled those systems. One of Perry's earliest projects involved intercepting the radio signals guiding a Soviet nuclear warhead as it plunged toward its target, then altering its trajectory, so the warhead swerved off course. Perry figured out a way to do this, but he told his bosses it wouldn't be of much use, since Soviet nuclear warheads were so powerful—several megatons of blast, to say nothing of thermal heat and radioactive fallout—that millions of Americans would die anyway. (This experience led Perry, years later, to become an outspoken advocate of nuclear arms-reduction treaties.)

Still, Perry grasped a key point that most other weapons scientists of the day did not: that getting inside the enemy's communications could drastically alter the effect of a weapon—and maybe the outcome of a battle or a war.

Perry rose through the ranks of Sylvania, taking over as director in 1954, then ten years later he left to form his own company, Electromagnetic Systems Laboratory, or ESL, which did contract work almost exclusively for the NSA and CIA. By the time he joined the Pentagon in 1977, he was as familiar as anyone with the spy agencies' advances in signals intelligence; his company, after all, had built the hardware that made most of those advances possible.

It was Perry who placed these scattershot advances under a single rubric: "counter-C2 warfare," the "C2" standing for "command and control." The phrase derived from his longtime preoccupation with electronic countermeasures, for instance jamming an enemy jet's radar receiver. But while jammers gave jets a *tactical* edge, counter-C2 warfare was a *strategic* concept; its goal was to degrade an enemy commander's ability to wage war. The concept regarded communications links—and the technology to intercept, disrupt, or sever them—not merely as a conveyor belt of warfare but as a decisive weapon in its own right.

When Jimmy Carter was briefed on these strategic breakthroughs, he seemed fascinated by the technology. When his successor, the Cold War hawk Ronald Reagan, heard the same briefing a year later, he evinced little interest in the technical details, but was riveted to the big picture: it meant that if war broke out between the superpowers, as many believed likely, the United States could win, maybe quickly and decisively.

In his second term as president, especially after the reformer Mikhail Gorbachev took over the Kremlin, Reagan rethought the implications of American superiority: he realized that his military's aggressive tactics and his own brazen rhetoric were making the Russians jumpy and the world more dangerous; so he softened his rhetoric, reached out to Gorbachev, and the two wound up signing a string of historic arms-reduction treaties that nearly brought the Soviet Union—the "evil empire," as Reagan had once described it—into the international order. But during his first term, Reagan pushed hard on his advantage, encouraging the NSA and other agencies to keep up the counter-C2 campaign.

Amid this pressure, the Russians didn't sit passive. When they found out about the microwaves emanating from the U.S. embassy's tenth floor, they started beaming its windows with their own microwave generators, hoping to listen in on the American spies' conversations.

The Russians grew clever at the spy-counterspy game. At one point, officials learned that the KGB was somehow stealing secrets from the Moscow embassy. The NSA sent over an analyst named Charles Gandy to solve the mystery. Gandy had a knack for finding trapdoors and vulnerabilities in any piece of hardware. He soon found a device called the Gunman inside sixteen IBM Selectric typewriters, which were used by the secretaries of high-level embassy officials. The Gunman recorded every one of their keystrokes and transmitted the data to a receiver in a church across the street. (Subsequent probes revealed that an attractive Russian spy had lured an embassy guard to let her in.)

It soon became clear that the Russians were setting up microwave beams and listening stations all over Washington, D.C., and New York City. Senior Pentagon officials—those whose windows faced high buildings across the Potomac River—took to playing Muzak in their offices while at work, so that if a Russian spy was shooting microwaves at those windows, it would clutter the ambient sound, drowning out their conversations.

Bobby Ray Inman had his aides assess the damage of this new form of spying. President Carter, a technically sophisticated engineer (he loved to examine the blueprints of the military's latest spy satellites), had been assured that his phone conversations, as well as those of the secretaries of state and defense, were carried on secure landlines. But NSA technicians traced those lines and discovered that, once the signal reached Maryland, it was shunted to microwave transmitters, which were vulnerable to interception. There was no evidence the Soviets *were* listening in, but there was no reason to think they weren't; they certainly *could* be, with little difficulty.

It took a while, but as more of these vulnerabilities were discovered, and as more evidence emerged that Soviet spies were exploiting them, a disturbing thought smacked a few analysts inside NSA: *Anything we're doing to them, they can do to us.*

This anxiety deepened as a growing number of corporations, public utilities, and government contractors started storing data and running operations on automated computers—especially since some of them were commingling classified and unclassified data on the same machines, even the same software. Willis Ware's warnings of a dozen years earlier were proving alarmingly prophetic.

Not everyone in the NSA was troubled. There was widespread complacency about the Soviet Union: doubt, even derision at the idea, that a country so technologically backward could do the remarkable things that America's SIGINT crews were doing. More than that, to the extent computer hardware and software had security holes, the NSA's managers were reluctant to patch them. Much of this hardware and software was used (or copied) in countries worldwide, including the targets of NSA surveillance; if it could easily be hacked, so much the better for surveillance.

The NSA had two main directorates: Signals Intelligence and Information Security (later called Information Assurance). SIGINT was the active, glamorous side of the puzzle palace: engineers, cryptologists, and old-school spies, scooping up radio transmissions, tapping into circuits and cables, all aimed at intercepting and analyzing communications that affected national security. Information Security, or INFOSEC, tested the reliability and security of the hardware and software that the SIGINT teams used. But for much of the agency's history, the two sides had no direct contact. They weren't even housed in the same building. Most of the NSA, including the SIGINT Directorate, worked in the massive complex at Fort Meade, Maryland. INFOSEC was a twenty-minute drive away, in a drab brown brick building called FANEX, an annex to Friendship Airport, which later became known as BWI Marshall Airport. (Until 1968, INFOSEC had been still more remote, in a tucked-away building—which, many years later, became the Department of Homeland Security headquarters—on Nebraska Avenue, in Northwest Washington.) INFOSEC technicians had a maintenance function; they weren't integrated into operations at all. And the SIGINT teams did nothing *but* operations; they didn't share their talents or insights to help repair the flaws in the equipment they were monitoring.

These two entities began to join forces, just a little, toward the end of Carter's presidency. Pentagon officials, increasingly aware that the Soviets were penetrating their communications links, wanted INFOSEC to start testing hardware and software used not only by the NSA but by the Defense Department broadly. Inman set up a new organization, called the Computer Security Center, and asked his science and technology chief, George Cotter, to direct it. Cotter was one of the nation's top cryptologists; he'd been doing signals intelligence since the end of World War II and had worked for the NSA from its inception. Inman wanted the new center to start bringing together the SIGINT operators and the INFOSEC technicians on joint projects. The cultures would remain distinct for years to come, but the walls began to give.

The order to create the Computer Security Center came from the ASD(C3I), the assistant secretary of defense for command, control, communications, and intelligence—the Pentagon's liaison with the NSA. When Reagan became president, his defense secretary, Caspar Weinberger, appointed Donald Latham to the position. Latham had worked SIGINT projects with George Cotter in the early to mid-1970s on the front lines of the Cold War: Latham as chief scientist of U.S. European Command, Cotter as deputy chief of NSA-Europe. They knew, as intimately as anyone, just how deeply both sides—the Soviets and the Americans (and some of their European allies, too)—were getting inside each other's communications channels. After leaving NSA, Latham was named deputy chief of the Pentagon's Office of Microwave, Space and Mobile Systems—and, from there, went on to work in senior engineering posts at Martin Marietta and RCA, where he remained immersed in these issues.

When General Jack Vessey came back from that White House meeting after Ronald Reagan had watched *WarGames* and asked his aides to find out whether someone could hack into the military's most sensitive computers, it was only natural that his staff would forward the question to Don Latham. It didn't take long for Latham to send back a response, the same response that Vessey would deliver to the president: *Yes, the problem is much worse than you think.*

Latham was put in charge of working up, and eventually drafting, the presidential directive called NSDD-145. He knew the various ways that the NSA—and, among all federal agencies, only the NSA—could not only hack but also secure telecommunications and computers. So in his draft, he put the NSA in charge of all their security.

The directive called for the creation of a National Telecommunications and Information Systems Security Committee "to consider technical matters" and "develop operating policies" for implementing the new policy. The committee's chairman would be the ASD(C3I)—that is to say, the chairman would be Don Latham.

The directive also stated that residing within this committee would be a "permanent secretariat composed of personnel of the National Security Agency," which "shall provide facilities and support as required." There would also be a "National Manager for Telecommunications and Automated Information Systems Security," who would "review and approve all standards, techniques, systems, and equipments." The directive specified that this National Manager would be the NSA director.

It was an ambitious agenda, too ambitious for some. Congressman Jack Brooks, a Texas Democrat and Capitol Hill's leading civil-liberties advocate, wasn't about to let the NSA—which was limited, by charter, to surveillance of foreigners—play any role in the daily lives of Americans. He wrote, and his fellow lawmakers passed, a bill that revised the president's directive and denied the agency any such power. Had Don Latham's language been left standing, the security standards and compliance of every computer in America—government, business, and personal—would have been placed under the tireless gaze of the NSA.

It wouldn't be the last time that the agency tried to assert this power—or that someone else pushed back.

CHAPTER 2

"IT'S ALL ABOUT THE INFORMATION"

O N August 2, 1990, Saddam Hussein, the president of Iraq, ordered his army to invade Kuwait, the small country to the south. Three days later, President George H. W. Bush declared that this aggression "will not stand." On January 17, 1991, after a massive mobilization, U.S. helicopters and combat planes fired the first shots of a month-long air campaign over Iraq—followed, on February 24, by a hundred-hour ground assault, involving more than a half million American troops enveloping and crushing the Iraqi army, pushing its scattered survivors back across the border.

Known as Operation Desert Storm, it was the largest armored offensive the world had seen since the Second World War. It was also—though few were aware of this—the first campaign of "counter command-control warfare," the harbinger of cyber wars to come.

The director of the NSA at the time was Rear Admiral William Studeman, who, like his mentor, Bobby Ray Inman, had been director of naval intelligence before taking the helm at Fort Meade. When Studeman was appointed to run the NSA, he took with him,

as his executive assistant, a veteran Navy cryptologist named Richard Wilhelm, who, a few years earlier, had been the number two at
the agency's large SIGINT site in Edsall, Scotland, running the test
bed for Inman's Bauded Signals Upgrade program, which aimed to
decrypt Soviet communications.

As the planning for Desert Storm got under way, Studeman sent
Wilhelm to the Pentagon as the NSA delegate to a hastily improvised
group called the Joint Intelligence Center. The head of the center
was Rear Admiral John "Mike" McConnell, who held the post of
J-2, the intelligence officer for the chairman of the Joint Chiefs of
Staff. Like most fast-rising officers in naval intelligence, Wilhelm
and McConnell had known each other for years. In the new center,
they created a multiservice apparatus that combined SIGINT, satellite imagery, and human spies on the ground into a single cell of
intelligence-gathering and -analysis.

Before the invasion, American intelligence officers knew little
about Iraq or Saddam Hussein's military machine. By the time the
bombing began, they knew most of what there was to know. Months
before the first shot was fired, McConnell's analysts penetrated deep
inside Saddam's command-control network. A key discovery was that
Saddam had run fiber-optic cable all the way from Baghdad down
to Basra and, after his invasion, into Kuwait City. American intel
officers contacted the Western firms that had installed the cable and
learned from them the locations of the switching systems. When the
bombing began in the wee hours of January 17, those switches were
among the first targets hit. Saddam had to reroute communications
to his backup network, built on microwave signals. Anticipating this
move, the NSA had positioned a new top secret satellite directly over
Iraq, one of three spy-in-the-sky systems that Wilhelm had managed
before the war. This one sported a receiver that scooped up microwave signals.

At every step, then, the NSA, McConnell's Joint Intelligence Center, and, through them, the American combat commanders knew exactly what Saddam and his generals were saying and where their soldiers were moving. As a result, the United States gained a huge edge in the fight: not only could its commanders swiftly counter the Iraqi army's moves, they could also move their own forces around without fear of detection. The Iraqis had lots of antiaircraft missiles, which they'd acquired over the years from the Soviets, and they were well trained to use them. They might have shot down more American combat planes, but McConnell's center figured out how to disrupt Iraq's command-control systems and its air-defense radar.

Saddam's intelligence officers soon detected the breach, so he started to send orders to the front via couriers on motorbikes; but this was a slow process, the war by then was moving fast, and there was little he could do to avoid a rout.

This first experiment in counter-C2 warfare was a success, but it didn't go very far, not nearly as far as its partisans could have taken it, because the U.S. Army's senior officers weren't interested. General Norman Schwarzkopf, the swaggering commander of Desert Storm, was especially dismissive. "Stormin' Norman" was old-school: wars were won by killing the enemy and destroying his targets, and, in this regard, all wars were the same: big or small, conventional or guerrilla, on the rolling hills of Europe, in the jungles of Vietnam, or across the deserts of Mesopotamia.

Initially, Schwarzkopf wanted nothing to do with the feeds from McConnell's center. He'd brought with him only a handful of intelligence officers, figuring they'd be sufficient for the job. The entire intelligence community—the directors of the CIA, NSA, Defense Intelligence Agency, and others—raised a fuss. It took the intervention of the chairman of the Joint Chiefs of Staff, Colin Powell, an Army general with Washington grooming and a strategic outlook, to bring the center's intel analysts into a conversation with the war planners.

Even so, Schwarzkopf put up resistance. When he learned that Saddam was transmitting his orders through microwaves after the fiber-optic cables were destroyed, his first instinct was to blow up the microwave link. Some of his own intel analysts argued against him, pointing to the reams of valuable information they were getting from the intercept. Schwarzkopf dismissed these objections, insisting that destroying Saddam's communication links, rather than exploiting them for intelligence, would be the speedier path to victory.

It wasn't just Schwarzkopf who waved away the Joint Intelligence Center's schemes; the Pentagon's top civilians were also leery. This was all very new. Few politicians or senior officials were versed in technology; neither President Bush nor his secretary of defense, Dick Cheney, had ever used a computer. At a crucial point in the war, as the American ground forces made their end run to attack the Iraqi army from the flanks and the rear, the NSA and the Joint Intelligence Center proposed disabling an Iraqi telecommunications tower by hacking into its electronics: the tower needed to be put out of action for just twenty-four hours; there was no need to blow it up (and probably kill some innocent people besides). Cheney was skeptical. He asked the analysts how confident they were that the plan would work; they were unable to quantify the odds. By contrast, a few bombs dropped from fighter planes would do the job with certainty. Cheney went with the bombs.

———

Those who were immersed in the secret counter-C2 side of the war came away feeling triumphant, but some were also perplexed and disturbed. Richard H. L. Marshall was a legal counselor for the NSA. Before the fighting started, he'd voiced some concerns about the battle plan. At one point, an Iraqi generator, which powered a military facility, was supposed to be disabled by electronic means. But Marshall saw that it also powered a nearby hospital. There was a chance that this attack—though it didn't involve bullets, missiles, or bombs—would nonetheless kill a lot of civilians, and the most helpless civilians at that.

Marshall and other lawyers, in the NSA and the Pentagon, held a spirited discussion about the implications. Their concerns proved moot: Schwarzkopf and other commanders decided to drop bombs and missiles on the generator and almost every other urban target— power plants, water-purification centers, communications towers, and various facilities having dual civil and military functions—and the "collateral damage" killed thousands of Iraqi civilians.

Still, from his vantage at NSA, Marshall could anticipate a growth spurt in this new sort of warfare—perhaps a time, in the not too distant future, when it matured to a dominant form of warfare. If a nation destroyed or disabled a piece of critical infrastructure, without launching a missile or dropping a bomb, would that constitute an act of war? Would its commanders and combatants be subject to the Law of Armed Conflict? Nobody knew; nobody with the authority to mull such matters had given it any thought.

Other NSA officers, more highly ranked and operationally oriented, had a different, more strategic concern. They were astonished by how easy it had been to take out Saddam's communications links. But some knew that, in a future war, especially against a foe more advanced than Iraq, it might not be so easy. The technology was changing: from analog to digital, from radio transmissions and microwaves to fiber optics, from discrete circuits of phone lines to data packets of what would come to be called cyberspace. Even Saddam Hussein had fiber-optic cable. Because European allies had installed it, American officials could learn where the switches were located and, therefore, where to drop the bombs. But one could imagine another hostile nation laying cable on its own. Or, if a war wasn't going on, if the NSA simply wanted to intercept signals whooshing through the cable, just as it had long been intercepting phone calls and radio transmissions, there would be no way to get inside. It might be technically possible to tap into the cable, but the NSA wasn't set up for the task.

The official most deeply worried about these trends was the NSA director, Bill Studeman.

In August 1988, a few days before Studeman took command at Fort Meade, Inman invited him and another old colleague, Richard Haver, to dinner. Seven years had passed since Inman had run the NSA, and he wasn't pleased with what his two successors—Lincoln Faurer and William Odom—had done with the place. Both were three-star generals, Faurer with the Air Force, Odom with the Army (the directorship usually rotated among the services), and to Inman, the career Navy man, that was part of the problem.

Of the military's three main services, the Navy was most attuned to shifts in surveillance technology. Its number-one mission was keeping track of the Soviet navy, especially Soviet submarines; and the U.S. Navy's most secretive branches conducted this hunt with many of the same tools and techniques that the NSA used. There was an esprit de corps among the coterie of Navy officers who rose through the ranks in these beyond-top-secret programs. In part, this was because they *were* so highly classified; having the clearances to know the slightest details about them made them members of the military's most secret club. In part, it stemmed from the intensity of their mission: what they did, 24/7, in peacetime—cracking Soviet codes, chasing Soviet submarines—was pretty much the same things they would do in wartime; the sense of urgency never let up.

Finally, this esprit had been the willed creation of Bobby Ray Inman. When he was director of naval intelligence in the mid-1970s, his top aides helped him identify the smartest people in the various branches of the Navy—attachés, officers on aircraft carriers, as well as black-ops submariners and cryptographers—and put them together in teams, to make sure that the most important intelligence got into the operators' hands and that the operators aligned their missions to the intel officers' needs.

Inman was a ruthless player of bureaucratic politics; Bill Studeman and Rich Haver liked to say that Machiavelli was an angel by comparison. As NSA director in the late 1970s and early 1980s, Inman engaged in protracted power struggles over which agency, NSA or CIA, would win control of the new technologies. When Ronald Reagan was elected president, he asked Inman, whose term as NSA director had nearly expired, to move over to Langley and become deputy director of the CIA. The Senate confirmed his nomination to the new job on February 12, 1981, but he remained director of the NSA until March 30. In that five-week period of dual powers, Inman sent several memos to himself—NSA director to CIA deputy director, and vice versa—and thereby settled many of the scores between the two agencies. (Inman's boss at the CIA, the director, William Casey, was focused more on secret wars against communists in Central America and Afghanistan, so didn't concern himself with internal matters.) In the end, the NSA was given sole control of computer-based intelligence. (This set the stage, three years later, for Reagan's NSDD-145, which, until Congress overrode it, gave the NSA the power to establish security standards for all telecommunications and computers; Inman's self-addressed memos had established the precedents for this authority.) In a few other disputes, Inman split the responsibilities, creating joint CIA-NSA teams. With the roles and missions secure, Inman also boosted both agencies' budgets for expensive hardware that he'd desired back at Fort Meade—including supercomputers and miniaturized chips that enhanced the collection powers of sensors on satellites, spy planes, and submarines.

Inman stayed at the CIA for less than two years, then retired from government, moved back to his native Texas, and made a fortune in start-up software and commercial-encryption companies. From that vantage, he saw how quickly the digital revolution was spreading worldwide—and how radically the NSA would have to change to keep pace. He remained active on government advisory boards, occasionally checked in with former underlings at Fort Meade, and grew frustrated that Linc Faurer, then Bill Odom, weren't paying attention to the sharp turns ahead.

Now, in the final year of Reagan's presidency, Bill Studeman—not only a Navy man with experience in classified projects, but also a fellow Texan and one of Inman's top protégés—was about to become the director of NSA. Rich Haver, who joined the two for dinner that summer night, was the deputy director of naval intelligence.

When Inman had been director of naval intelligence, Studeman and Haver had worked on his staff. Studeman had worked on the advances in surveillance and computer processing, including the Bauded Signals Upgrade, that gave America a leg up on Russia at the beginning of Reagan's presidency. Haver, a persuasive figure with a slide show and a pointer, was the one who briefed the president and his top aides on the advances' implications. The three of them—Inman, Studeman, and Haver—had degrees in history, not physics or engineering. The world was changing; the Cold War was entering a new phase; and they saw themselves as frontline players in a realm of the struggle that almost no one else knew existed.

Inman called together his two former underlings that night to tell them—really, to lecture them, through the entirety of a three-hour dinner—that they had to push the intelligence community, especially the NSA, out in front of the technological changes. They had to alter the way the agencies did business, promoted their personnel, and focused their energies.

Among the first things that Studeman did when he assumed the helm at Fort Meade a few days later, was to commission two papers. One, called the "Global Access Study," projected how quickly the world would shift from analog to digital. It concluded that the change wouldn't take place all at once or uniformly; that the NSA would have to innovate in order to meet the demands (and intercept the communications) of the new world, while still monitoring the present landscape of telephone, radio, and microwave signals.

Studeman's second paper, an analysis of NSA personnel and their skill sets, concluded that the balance was wrong: there were too many Kremlinologists, not enough computer scientists. When Inman was director, he'd taken a few small steps to bring the technicians into the same room as the SIGINT operators and analysts, but the effort had since stalled. Most of the agency's computer experts worked in IT or maintenance. No one in SIGINT was tapping their expertise for advice on vulnerabilities in new hardware and software. In short, no one was preparing for the new era.

Studeman's studies—the very fact that he commissioned them—sparked resistance, anger, and fear from the rank and file. Over the years, the NSA's managers had invested, and were still spending, colossal sums on analog technology, and they chose to ignore or dismiss warnings that they'd made a bad bet. The old guard took Studeman's second study—the one on the looming mismatch between the agency's skill sets and its mission—as a particularly ominous threat: if the new director acted on his study's conclusions, thousands of veteran analysts and spies would soon be out of a job.

There was only so much Studeman could do during his three years in charge. For one thing, the world was changing more quickly than anyone could have imagined. By the time Studeman left Fort Meade in April 1992, the Cold War—the struggle that had animated the NSA since its birth—was over and won. Even if the need for NSA reform had been widely accepted (and it wasn't), it suddenly seemed less urgent.

———

Studeman's successor was Rear Admiral Mike McConnell, who had run the Joint Intelligence Center during Operation Desert Storm.

McConnell had remained General Powell's intelligence officer in the year and a half since the war. In the mid-1980s, he'd spent a year-long tour at NSA headquarters, attached to the unit tracking Soviet naval forces. But returning to Fort Meade as NSA director, at a moment of such stark transition, McConnell didn't quite know what he and this enormous agency were supposed to do.

There were two distinct branches of the agency's SIGINT Directorate: the "A Group," which monitored the Soviet Union and its satellites; and the "B Group," which monitored the rest of the world. As its title suggested, the A Group was the elite branch, and everyone in the building knew it. Its denizens imbibed a rarefied air: *they* were the ones protecting the nation from the rival superpower; they had learned the imponderably specialized skills, and had immersed themselves so deeply into the Soviet mindset, that they could take a stream of seemingly random data and extract patterns and shifts of patterns that, pieced together, gave them (at least in theory) a picture of the Kremlin's intentions and the outlook for war and peace. Now that the Cold War was over, what good were those skills? Should the Kremlin-watchers still be called the A Group?

A still larger uncertainty was how the NSA, as a whole, would continue to do its watching—and listening. Weeks into his tenure as director, McConnell learned that some of the radio receivers and antennas, which the NSA had set up around the globe, were no longer picking up signals. Studeman's "Global Access Study"—which predicted the rate at which the world would switch to digital—was coming true.

Around the same time, one of McConnell's aides came into his office with two maps. The first was a standard map of the world, with arrows marking the routes that the major shipping powers navigated across the oceans—the "sea lines of communication," or SLOCs, as a Navy man like McConnell would have called them. The second map showed the lines and densities of fiber-optic cable around the world.

This is the map that you should study, the aide said, pointing to the second one. Fiber-optic lines were the new SLOCs, but they were to SLOCs what wormholes were to the galaxies: they whooshed you from one point to any other point *instantaneously*.

McConnell got the parallel, and the hint of transformation, but he didn't quite grasp its implications for his agency's future.

Shortly after that briefing, he saw a new movie called *Sneakers*. It was a slick production, a comedy-thriller with an all-star cast. The only reason McConnell bothered to see the film was that someone had told him it was about the NSA. The plot was dopey: a small company that does white-hat hacking and high-tech sleuthing is hired to steal a black box sitting on a foreign scientist's desk; the clients say that they're with the NSA and that the scientist is a spy; as it turns out, the clients are spies, the scientist is an agency contractor, the black box is a top secret device that can decode all encrypted data, and the NSA wants it back; the sleuths are on the case.

Toward the end of the film, there was a scene where the evil genius (played by Ben Kingsley), a former computer-hacking prankster who turns out to have ordered the theft of the black box, confronts the head sleuth (played by Robert Redford), an old friend and erstwhile comrade from their mischievous college days, and uncorks a dark soliloquy, explaining why he stole the box:

"The world isn't run by weapons anymore, or energy, or money," the Kingsley character says at a frenzied clip. "It's run by ones and zeroes, little bits of data. It's all just electrons. . . . There's a war out there, old friend, a world war. And it's not about who's got the most bullets. It's about who controls the information: what we see and hear, how we work, what we think. It's all about the information."

McConnell sat up as he watched this scene. Here, in the unlikely form of a Hollywood movie, was the NSA mission statement that he'd been seeking: *The world is run by ones and zeroes . . . There's a war out there . . . It's about who controls the information.*

Back at Fort Meade, McConnell spread the word about *Sneakers*, encouraged every employee he ran into to go see it. He even obtained a copy of the final reel and screened it for the agency's top officials, telling them that this was the vision of the future that they should keep foremost in their minds.

He didn't know it at the time, but the screenplay for *Sneakers* was cowritten by Larry Lasker and Walter Parkes—the same pair that, a decade earlier, had written *WarGames*. And, though not quite to the same degree, *Sneakers,* too, would have an impact on national policy.

Soon after his film-inspired epiphany, McConnell called Rich Wilhelm, who'd been the NSA representative—in effect, his right-hand man—on the Joint Intelligence Center during Desert Storm. After the war, Wilhelm and Rich Haver had written a report, summarizing the center's activities and listing the lessons learned for future SIGINT operations. As a reward, Wilhelm was swiftly promoted to take command of the NSA listening station at Misawa Air Base in Japan, one of the agency's largest foreign sites. In the order of NSA field officers, Wilhelm was king of the hill.

But now, McConnell was asking Wilhelm to come back to Fort Meade and take on a new job that he was creating just for him. Its title would be Director of Information Warfare. (*There's a war out there . . . It's about who controls the information.*)

The concept, and the nomenclature, spread. The following March, General Colin Powell, chairman of the Joint Chiefs of Staff, issued a policy memorandum on "information warfare," which he defined as operations to "decapitate the enemy's command structure from its body of combat forces." The military services responded almost at once, establishing the Air Force Information Warfare Center, the Naval Information Warfare Activity, and the Army Land Information Warfare Activity. (These entities already existed, but under different names.)

By the time McConnell watched *Sneakers*, he'd been fully briefed on the Navy and NSA programs in counter-C2 warfare, and he was intrigued with the possibilities of applying the concept to the new era. In its modern incarnation ("information warfare" was basically counter-C2 warfare plus digital technology), he could turn SIGINT on its ear, not just intercepting a signal but penetrating its source—and, once inside the mother ship, the enemy's command-control system, he could feed it false information, altering, disrupting, or destroying the machine, disorienting the commanders: *controlling* the information to keep the peace and win the war.

None of this came as news to Wilhelm; he'd been skirmishing on the information war's front lines for years. But six weeks into the new job, he came to McConnell's office and said, "Mike, we're kind of fucked here."

Wilhelm had been delving into the details of what information *war*—a *two-way* war, in which both sides use the same weapons—might look like, and the sight wasn't pretty. The revolution in digital signals and microelectronics was permeating the American military and American society. In the name of efficiency, generals and CEOs alike were hooking up *everything* to computer networks. The United States was growing more dependent on these networks than any country on earth. About 90 percent of government files, including intelligence files, were flowing alongside commercial traffic. Banks, power grids, pipelines, the 911 emergency call system—all of these enterprises were controlled through networks, and all of them were vulnerable, most of them to very simple hacking.

When you think about attacking someone else's networks, Wilhelm told McConnell, keep in mind that *they* can do the same things to *us*. Information warfare wasn't just about gaining an advantage in combat; it also had to be about protecting the nation from other countries' efforts to gain the same advantage.

It was a rediscovery of Willis Ware's warning from a quarter century earlier.

McConnell instantly grasped the importance of Wilhelm's message. The Computer Security Center, which Bobby Ray Inman created a decade earlier, had since lured little in the way of funding or attention. The Information Security (now called Information Assurance) Directorate was still—literally—a sideshow, located a twenty-minute drive from headquarters.

Meanwhile, the legacy of Reagan's presidential directive on computer security, NSDD-145, lay in tatters. Congressman Jack Brooks's overhaul of the directive, laid out in the Computer Security Act of 1987, gave NSA control over the security of *military* computers and *classified* networks, but directed the National Bureau of Standards, under the Department of Commerce, to handle the rest. The formula was doomed from the start: the NBS lacked technical competence, while the NSA lacked institutional desire. When someone at the agency's Information Assurance Directorate or Computer Security Center discovered a flaw in a software program that another country might also be using, the real powers at NSA—the analysts in the SIGINT Directorate—wanted to exploit it, not fix it; they saw it as a new way to penetrate a foreign nation's network and to intercept its communications.

In other words, it wasn't so much that the problem went ignored; rather, no one in power saw it as a problem.

McConnell set out to change that. He elevated the Information Assurance Directorate, gave it more money at a time when the overall budget—not just for the NSA but for the entire Defense Department—was getting slashed, and started moving personnel back and forth, between the SIGINT and Information Assurance directorates, just for short-term tasks, but the idea was to expose the two cultures to one another.

It was a start, but not much more than that. McConnell had a lot on his plate: the budget cuts, the accelerating shift from analog circuits to digital packets, the drastic decline in radio signals, and the resulting need to find new ways to intercept communications. (Not long after McConnell became director, he found himself having to shut down one of the NSA antennas in Asia; it was picking up *no* radio signals; *all* the traffic that it had once monitored, in massive volume at its peak, had moved to underground cables or cyberspace.)

In the fall of 1994, McConnell saw a demonstration, in his office, of the Netscape Matrix—one of the first commercial computer network browsers. He thought, "This is going to change the world." *Everyone* was going to have access to the Net—not just allied and rival governments, but individuals, including terrorists. (The first World Trade Center bombing had taken place the year before; terrorism, seen as a nuisance during the nuclear arms race and the Cold War, was emerging as a major threat.) With the rise of the Internet came commercial encryption, to keep network communications at least somewhat secure. Code-making was no longer the exclusive province of the NSA and its counterparts; everyone was doing it, including private firms in Silicon Valley and along Route 128 near Boston, which were approaching the agency's technical prowess. McConnell feared that the NSA would lose its unique luster—its ability to tap into communications affecting national security.

He was also coming to realize that the agency was ill equipped to seize the coming changes. A young man named Christopher Mellon, on the Senate Intelligence Committee's staff, kept coming around, asking questions. Mellon had heard the briefings on Fort Meade's adaptations to the new digital world; but when he came to headquarters and examined the books, he discovered that, of the agency's $4 billion budget, just $2 *million* was earmarked for programs to penetrate communications on the Internet. Mellon asked to see the personnel assigned to this program; he was taken to a remote corner of the main floor, where a couple dozen techies—out of a workforce numbered in the tens of thousands—were fiddling with computers.

McConnell hadn't known just how skimpy these efforts were, and he assured the Senate committee that he would beef up the programs as a top priority. But he was diverted by what he saw as a more urgent problem—the rise of commercial *voice* encryption, which would soon make it very difficult for the NSA (and the FBI) to tap phone conversations. McConnell's staff devised what they saw as a solution to the problem—the Clipper Chip, an encryption key that they billed as perfectly secure. The idea was to install the chip in every telecommunications device. The government could tap in and listen to a phone conversation, only if it followed an elaborate, two-key procedure. An agent would have to go to the National Institute of Standards and Technology, as the National Bureau of Standards was now called, to get one of the crypto-keys, stored on a floppy disk; another agent would go to the Treasury Department to get the other key; then the two agents would go to the Marine base at Quantico, Virginia, to insert both disks into a computer, which would unlock the encryption.

McConnell pushed hard for the Clipper Chip—he made it his top priority—but it was doomed from the start. First, it was expensive: a phone with a Clipper Chip installed would cost more than a thousand dollars. Second, the two-key procedure was baroque. (Dorothy Denning, one of the country's top cryptologists, took part in a simulated exercise. She obtained the key from Treasury, but after driving out to Quantico, learned that the person from NIST had picked up the wrong key. They couldn't unlock the encryption.) Finally, there was the biggest obstacle: very few people trusted the Clipper Chip, because very few people trusted the intelligence agencies. The revelations of CIA and NSA domestic surveillance, unleashed by Senator Frank Church's committee in the mid-1970s, were still a fresh memory. Nearly everyone—even those who weren't inclined to distrust spy agencies—suspected that the NSA had programmed the Clipper Chip with a *secret* back door that its agents could open, then listen to phone conversations, without going through Treasury, NIST, or any legal process.

The Clipper Chip ended with a whimper. It was McConnell's well-intentioned, but misguided, effort to forge a compromise between personal privacy and national security—and to do so openly, in the public eye. The next time the NSA created or discovered back doors into data, it would do so, as it had always done, under the cloak of secrecy.

CHAPTER 3

A CYBER PEARL HARBOR

ON April 19, 1995, a small gang of militant anarchists, led by Timothy McVeigh, blew up a federal office building in Oklahoma City, killing 168 people, injuring 600 more, and destroying or damaging 325 buildings across a sixteen-block radius, causing more than $600 million in damage. The shocking thing that emerged from the subsequent investigation was just how easily McVeigh and his associates had pulled off the bombing. It took little more than a truck and a few dozen bags of ammonium nitrate, a common chemical in fertilizers, obtainable in many supply stores. Security around the building was practically nonexistent.

The obvious question, in and out of the government, was what sorts of targets would get blown up next: a dam, a major port, the Federal Reserve, a nuclear power plant? The damage from any of those hits would be more than simply tragic; it could reverberate through the entire economy. So how vulnerable were they, and what could be done to protect them?

On June 21, Bill Clinton signed a Presidential Decision Directive, PDD-39, titled "U.S. Policy on Counterterrorism," which, among other things, put Attorney General Janet Reno in charge of a "cabinet

committee" to review—and suggest ways to reduce—the vulnerability of "government facilities" and "critical national infrastructure."

Reno turned the task over to her deputy, Jamie Gorelick, who set
up a Critical Infrastructure Working Group, consisting of other deputies from the Pentagon, CIA, FBI, and the White House. After a few
weeks of meetings, the group recommended that the president appoint a commission, which in turn held hearings and wrote a report,
which culminated in the drafting of another presidential directive.

Several White House aides, who figured the commission would
come up with new ways to secure important physical structures,
were startled when more than half of its report and recommendations dealt with the vulnerability of computer networks and the urgent need for what it called "cyber security."

The surprise twist came about because key members of the Critical Infrastructure Working Group and the subsequent presidential
commission had come from the NSA or the Navy's super-secret
black programs and were thus well aware of this new aspect of the
world.

Rich Wilhelm, the NSA director of information warfare, was
among the most influential members of the working group. A few
months before the Oklahoma City bombing, President Clinton had
put Vice President Al Gore in charge of overseeing the Clipper Chip;
Mike McConnell sent Wilhelm to the White House as the NSA liaison on the project. The chip soon died, but Gore held on to Wilhelm
and made him his intelligence adviser, with a spot on the National
Security Council staff. Early on at his new job, Wilhelm told some
of his fellow staffers about the discoveries he'd made at Fort Meade,
especially those highlighting the vulnerability of America's increasingly networked society. He wrote a memo on the subject for Clinton's national security adviser, Anthony Lake, who signed it with his
own name and passed it on to the president.

When Jamie Gorelick put together her working group, it was natural that Wilhelm would be on it. One of its first tasks was to define its title, to figure out which infrastructures were *critical*—which sectors were vital to the functioning of a modern society. The group came up with a list of eight: telecommunications, electrical power, gas and oil, banking and finance, transportation, water supply, emergency services, and "continuation of government" in the event of war or catastrophe.

Wilhelm pointed out that all of these sectors relied, in some cases heavily, on computer networks. Terrorists wouldn't need to blow up a bank or a rail line or a power grid; they could merely disrupt the computer network that controlled its workings, and the result would be the same. As a result, Wilhelm argued, "critical infrastructure" should include not just physical buildings but the stuff of what would soon be called cyberspace.

Gorelick needed no persuading on this point. As deputy attorney general, she served on several interagency panels, one of which dealt with national security matters. She co-chaired that panel with the deputy director of the CIA, who happened to be Bill Studeman, the former NSA director (and Bobby Ray Inman protégé). In his days at Fort Meade, Studeman had been a sharp advocate of counter-C2 warfare; at Langley he was promoting the same idea, now known as information warfare, both its offensive and its defensive sides—America's ability to penetrate the enemy's networks and the enemy's ability to penetrate America's.

Studeman and Gorelick met to discuss these issues every two weeks, and his arguments had resonance. Before her appointment as deputy attorney general, Gorelick had been general counsel at the Pentagon, where she heard frequent briefings on hackings of defense contractors and even of the Defense Department. Now, at the Justice Department, she was helping to prosecute criminal cases of hackers who'd penetrated the computers of banks and manufacturers. One year before Oklahoma City, Gorelick had helped draft the Computer

Crime Initiative Action Plan, to boost the Justice Department's expertise in "high-tech matters," and had helped create the Information Infrastructure Task Force Coordinating Committee.

These ventures weren't mere hobbyhorses; they were mandated by the Justice Department's caseload. In recent times, a Russian crime syndicate had hacked into Citibank's computers and stolen $10 million, funneling it to separate accounts in California, Germany, Finland, and Israel. A disgruntled ex-employee of an emergency alert network, covering twenty-two states, crashed the system for ten hours. A man in California gained control of the computer running local phone switches, downloaded information about government wiretaps on suspected terrorists, and posted the information online. Two teenage boys, malicious counterparts to the hero of *WarGames*, hacked into the computer network at an Air Force base in Rome, New York; one of the boys later sneered that military sites were the easiest to hack on the entire Internet.

From all this—her experiences as a government lawyer, the interagency meetings with Studeman, and now the discussions with Rich Wilhelm on the working group—Gorelick was coming to two disturbing conclusions. First, at least in this realm, the threats from criminals, terrorists, and foreign adversaries were all the same: they used the same means of attack; often, they couldn't be distinguished. This wasn't a problem for the Department of Justice or Defense alone; the whole government had to deal with it, and, since most computer traffic moved along networks owned by corporations, the private sector had to help find, and enforce, solutions, too.

Second, the threat was wider and deeper than she'd imagined. Looking over the group's list of "critical" infrastructures, and learning that they were all increasingly controlled by computers, Gorelick realized, in a jaw-drop moment, that a coordinated attack by a handful of technical savants, from just down the street or the other side of the globe, could devastate the nation.

What nailed this new understanding was a briefing by the Pentagon's delegate to the working group, a retired Navy officer named Brenton Greene, who had recently been named to a new post, the director for infrastructure policy, in the office of the undersecretary of defense.

Greene had been involved in some of the military's most highly classified programs. In the late 1980s and early 1990s, he was a submarine skipper on beyond-top-secret spy missions. After that, he managed Pentagon black programs in a unit called the J Department, which developed emerging technologies that might give America an advantage in a coming war. One branch of J Department worked on "critical-node targeting." The idea was to analyze the infrastructures of every adversary's country and to identify the key targets—the smallest number of targets that the American military would have to destroy in order to make a huge impact on the course of a war. Greene helped to develop another branch of the department, the Strategic Leveraging Project, which focused on new ways of penetrating and subverting foreign adversaries' command-control networks—the essence of information warfare.

Working on these projects and seeing how easy it was, at least in theory, to devastate a foreign country with a few well-laid bombs or electronic intrusions, Greene realized—as had several others who'd journeyed down this path before him—the flip side of the equation: what we could do to them, they could do to us. And Greene was also learning that America was far more vulnerable to these sorts of attacks—especially information attacks—than any other country on the planet.

In the course of his research, Greene came across a 1990 study by the U.S. Office of Technology Assessment, a congressional advisory group, called *Physical Vulnerability of Electric Systems to Natural Disasters and Sabotage*. In its opening pages, the authors revealed which power stations and switches, if disabled, would take down huge chunks of the national grid. This was a public document, available to anyone who knew about it.

One of Greene's colleagues in the J Department told him that, soon after George Bush entered the White House in January 1989, Senator John Glenn showed the study to General Brent Scowcroft, Bush's national security adviser. Scowcroft was concerned and asked a Secret Service officer named Charles Lane to put together a small team—no more than a half dozen technical analysts—to do a separate study. The team's findings were so disturbing that Scowcroft shredded all of their work material. Only two copies of Lane's report were printed. Greene obtained one of them.

At this point, Greene concluded that he'd been working the wrong side of the problem: protecting America's infrastructure was more vital—as he saw it, more urgent—than seeking ways to poke holes in foreign infrastructures.

Greene knew Linton Wells, a fellow Navy officer with a deep background in black programs, who was now military assistant to Walter Slocombe, the undersecretary of defense for policy. Greene told Wells that Slocombe should hire a director for infrastructure policy. Slocombe approved the idea. Greene was hired.

In his first few months on the new job, Greene worked up a briefing on the "interdependence" of the nation's infrastructure, its concentration, and the commingling of one segment with the others—how disabling a few "critical nodes" (a phrase from J Department) could severely damage the country.

For instance, Greene knew that the Bell Corporation distributed a CD-ROM listing all of its communications switches worldwide, so that, say, a phone company in Argentina would know how to connect circuits for routing a call to Ohio. Greene looked at this guide with a different question in mind: where were all the switches in the major American cities? In each case he examined, the switches were—for reasons of economic efficiency—concentrated at just a couple of sites. For New York City, most of them were located at two addresses in Lower Manhattan: 140 West Street and 104 Broad Street. Take out those two addresses—whether with a bomb or an information warfare attack—and New York City would lose almost all of its phone service, at least for a while. The loss of phone service would affect other infrastructures, and on the cascading would go.

Capping Greene's briefing, the CIA—where Bill Studeman was briefly acting director—circulated a classified report on the vulnerability of SCADA systems. The acronym stood for Supervisory Control and Data Acquisition. Throughout the country, again for economic reasons, utility companies, waterworks, railway lines—vast sectors of critical infrastructure—were linking one local stretch of the sector to another, through computer networks, and controlling all of them remotely, sometimes with human monitors, often with automated sensors. Before the CIA report, few on the working group had ever heard of SCADA. Now, everyone realized that they were probably just scratching the surface of a new danger that came with the new technology.

Gorelick wrote a memo, alerting her superiors that the group was expanding the scope of its inquiry, "in light of the breadth of critical infrastructures and the multiplicity of sources and forms of attack." It was no longer enough to consider the likelihood and impact of terrorists blowing up critical buildings. The group—and, ultimately, the president—also had to consider "threats from other sources."

What to call these "other" threats? One word was floating around in stories about hackings of one sort or another: "cyber." The word had its roots in "cybernetics," a term dating back to the mid-nineteenth century, describing the closed loops of information systems. But in its present-day context of computer networks, the term stemmed from William Gibson's 1984 science-fiction novel, *Neuromancer*, a wild and eerily prescient tale of murder and mayhem in the virtual world of "cyberspace."

Michael Vatis, a Justice Department lawyer on the working group who had just read Gibson's novel, advocated the term's adoption. Others were opposed: it sounded too sci-fi, too frivolous. But once uttered, the word snugly fit. From that point on, the group—and others who studied the issue—would speak of "cyber crime," "cyber security," "cyber war."

What to do about these cyber threats? That was the real question, the group's raison d'être, and here they were stuck. There were too many issues, touching too many interests—bureaucratic, political, fiscal, and corporate—for an interagency working group to settle.

On February 6, 1996, Gorelick sent the group's report to Rand Beers, Clinton's intelligence adviser and the point of contact for all issues related to PDD-39, the presidential directive on counterterrorism policy, which had set this study in motion. The report's main point—noting the existence of two kinds of threats to critical infrastructure, physical and cyber—was novel, even historic. As for a plan of action, the group fell back on the usual punt by panels of this sort when they don't know what else to do: it recommended the creation of a presidential commission.

For a while, nothing happened. Rand Beers told Gorelick that her group's report was under consideration, but there was no follow-up. A spur was needed. She found it in the personage of Sam Nunn, the senior Democrat on the Senate Armed Services Committee.

Gorelick knew Nunn from her days as the Pentagon's general counsel. Both were Democratic hawks, not quite a rare breed but not so common either, and they enjoyed discussing the issues with each other. Gorelick told him about her group's findings. In response, Nunn inserted a clause in that year's defense authorization bill, requiring the executive branch to report to Congress on the policies and plans to ward off computer-based attacks against the national infrastructure.

Nunn also asked the General Accounting Office, the legislature's watchdog agency, to conduct a similar study. The resulting GAO report, "Information Security: Computer Attacks at Department of Defense Pose Increasing Risks," cited one estimate that the Defense Department "may have experienced as many as 250,000 attacks last year," two thirds of them successful, and that "the number of attacks is doubling each year, as Internet use increases along with the sophistication of 'hackers' and their tools."

Not only was this figure unlikely (a quarter million attacks a year meant 685 *per day*, with 457 actual penetrations), it was probably pulled out of a hat: as the GAO authors themselves acknowledged, only "a small portion" of attacks were "actually detected and reported."

Still, the study sent a shockwave through certain corridors. Gorelick made sure that Beers knew about the wave's reverberations and warned him that Nunn was about to hold hearings on the subject. The president, she hinted, would do well to get out in front of the storm.

Nunn scheduled his hearing for July 16. On July 15, Clinton issued Executive Order 13010, creating the blue-ribbon commission that Gorelick's working group had suggested. The order, a near-exact copy of the working group's proposed draft three months earlier, began: "Certain national infrastructures are so vital that their incapacity or destruction would have a debilitating impact on the defense or economic security of the United States." Listing the same eight "critical" sectors that the working group had itemized, the order went on, "Threats to these critical infrastructures fall into two categories: physical threats to tangible property ('physical threats') and threats of electronic, radio-frequency, or computer-based attacks on the information or communications components that control critical infrastructures ('cyber threats')."

The next day, the Senate Governmental Affairs Committee, where Nunn sat as a top Democrat, held its much-anticipated hearing on the subject. One of the witnesses was Jamie Gorelick, who warned, "We have not yet had a terrorist cyber attack on the infrastructure. But I think that that is just a matter of time. We do not want to wait for the cyber equivalent of Pearl Harbor."

The cyber age was officially under way.

———

So, behind the scenes, was the age of cyber warfare. At one meeting of the Critical Infrastructure Working Group, Rich Wilhelm took Jamie Gorelick aside and informed her, in broad terms, of the ultrasecret flip side of the threat she was probing—that *we* had long been doing to other countries what some of those countries, or certain people in those countries, were starting to do to us. We weren't robbing their banks or stealing their industrial secrets, we had no need to do that; but we were using cyber tools—"electronic, radio-frequency, or computer-based attacks," as Clinton's executive order would put it—to spy on them, scope out their networks, and prepare the battlefield to our advantage, should there someday be a war.

The important thing, Wilhelm stressed, was that *our* cyber offensive capabilities must be kept off the table—must not even be hinted at—when discussing our vulnerability to other countries' cyber offensive capabilities. America's programs in this realm were among the most tightly held secrets in the entire national security establishment.

When Rand Beers met with deputies from various cabinet departments to discuss Clinton's executive order, John White, the deputy secretary of defense, made the same point to his fellow deputy sec-

retaries, in the same solemn tone: no one can so much as mention
America's cyber offensive capabilities.

The need for secrecy wasn't the only reason for the ensuing si-
lence on the matter. No one around the table said so, but, clearly, to
acknowledge America's cyber prowess, while decrying the prowess
of others, would be awkward, to say the least.

————

It took seven months for the commission to get started. Beers, who
once again served as the White House point man, first had to find
a place for the commissioners to meet. The Old Executive Office
Building, the mansion next door to the White House, wasn't suffi-
ciently wired for computer connections (in itself, a commentary on
the dismal state of preparedness for a cyber crisis). John Deutch, the
new CIA director, pushed for the commissioners to work at his headquarters in Langley, where they could have secure access to anything
they needed; but officials in other departments feared this might breed
insularity and excessive dependence on the intelligence community.
In the end, Beers found a vacant suite of offices in a Pentagon-owned
building in Arlington; to sweeten the deal, the Defense Department
offered to pay all expenses and to offer technical support.

Then came the matter of filling the commission. This was a
delicate matter. Nearly all of the nation's computer traffic flowed
through networks owned by private corporations; those corporations should have a say in their fate. Beers and his staff listed the ten
federal departments and agencies that would be affected by whatever recommendations came out of this enterprise—Defense, Justice, Transportation, Treasury, Commerce, the Federal Emergency
Management Administration, the Federal Reserve, the FBI, the CIA,
and the NSA—and decided that each agency head would pick two
delegates for the commission: one official and one executive from a
private contractor. In addition to deputy assistant secretaries, there

would also be directors or technical vice presidents from the likes of AT&T, IBM, Pacific Gas & Electric, and the National Association of Regulatory Utility Commissioners.

There was another delicate matter. The commission's final report would be a public document, but its working papers and meetings would be classified; the commissioners would need to be vetted for top secret security clearances. That, too, would take time.

Finally, Beers and the cabinet deputies had to pick a chairman. There were tried-and-true criteria for such a post: he (and it was almost always a he) should be eminent, but not famous; somewhat familiar with the subject at hand, but not an expert; respected, amiable, but not flush with his own agenda; someone with time on his hands, but not a reject or a duffer. They came up with a retired Air Force four-star general named Robert T. Marsh.

Tom Marsh had risen through the ranks on the technical side of the Air Force, especially in electronic warfare. He wound up his career as commander of the electronic systems division at Hanscom Air Force Base in Massachusetts, then as commander of Air Force Systems Command at Andrews Air Force Base, near Washington. He was seventy-one years old; since retiring from active duty, he'd served on the Defense Science Board and the usual array of corporate boards; at the moment, he was director of the Air Force Aid Society, the service's main charity organization.

In short, he seemed ideal.

John White, the deputy secretary of defense, called Marsh to ask if he would be willing to serve the president as chairman of a commission to protect critical infrastructure. Marsh replied that he wasn't quite sure what "critical infrastructure" meant, but he'd be glad to help.

To prepare for the task, Marsh read the report by Gorelick's Critical Infrastructure Working Group. It rang true. He recalled his days at Hanscom in the late 1970s and early 1980s, when the Air Force crammed new technologies onto combat planes with no concern for the vulnerabilities they might be sowing. The upgrades were all dependent on command-control links, which had no built-in redundancies. A few technically astute junior officers on Marsh's staff warned him that, if the links were disrupted, the plane would be disabled, barely able to fly, much less fight.

Still, Marsh had been away from day-to-day operations for twelve years, and this focus on "cyber" was entirely new to him. For advice and a reality check, Marsh called an old colleague who knew more about these issues than just about anybody—Willis Ware.

Ware had kept up with every step of the Internet revolution since writing his seminal paper, nearly thirty years earlier, on the vulnerability of computer networks. He still worked at the RAND Corporation, and he was a member of the Air Force Scientific Advisory Board, which is where Marsh had come to know and trust him. Ware assured Marsh that Gorelick's report was on the right track; that this was a serious issue and growing more so by the day, as the military and society grew more dependent on these networks; and that too few people were paying attention.

His chat with Ware filled Marsh with confidence. The president's executive order had chartered the commission to examine vulnerabilities to physical threats and cyber threats. Marsh figured that solutions to the physical threats were fairly straightforward; the cyber threats were the novelty, so he would focus his inquiry on them.

Marsh and the commissioners first convened in February 1997. They had six months to write a report. A few of the members were holdovers from the Critical Infrastructure Working Group, most notably Brent Greene, the Pentagon delegate, whose briefing on the vulnerability of telecom switches and the electrical power grid had so shaken Gorelick and the others. (Gorelick, who left the Justice Department for a private law practice in May, would later co-chair an advisory panel for the commission, along with Sam Nunn.)

Most of the commissioners were new to the issues—at best, they knew a little bit about the vulnerabilities in their own narrow sectors, but had no idea of how vastly they extended across the economy—and their exposure to all the data, at briefings and hearings, filled them with dread and urgency.

Marsh's staff director was a retired Air Force officer named Phillip Lacombe, who'd earned high marks as chief of staff on a recent panel studying the roles and missions of the armed forces. Lacombe's cyber epiphany struck one morning, when he and Marsh were about to board an eight a.m. plane for Boston, where they were scheduled to hold a ten-thirty hearing. Their flight was delayed for three hours because the airline's computer system was down; the crew couldn't measure weights and balances (a task once performed with a slide rule, which no one knew how to use anymore), so the plane couldn't take off. The irony was overwhelming: here they were, about to go hear testimony on the nation's growing dependence on computer networks—and they couldn't get there on time because of the nation's growing dependence on computer networks.

That's when Lacombe first realized that the problem extended to every corner of modern life. Military officers and defense intellectuals had been worried about weapons of mass destruction; Lacombe now saw there were also weapons of mass *disruption*.

Nearly every hearing that the commission held, as well as several casual conversations before and after, hammered home the same point. The executives of Walmart told the commission that, on a recent Sunday, the company's computer system crashed and, as a result, they couldn't open any of their retail stores in the southeast region of the United States. When a director at Pacific Gas & Electric, one of the nation's largest utilities, testified that all of its control systems were getting hooked up to the Internet, to save money and speed up the transmission of energy, Lacombe asked what the company was doing about security. He didn't know what Lacombe was talking about. Various commissioners asked the heads of railways and airlines how they were assuring the security of computer-controlled switches, tracks, schedules, and air traffic radar—and it was the same story: the corporate heads looked puzzled; they had no idea that security was an issue.

On October 13, 1997, the President's Commission on Critical Infrastructure Protection, as it was formally called, released its report—154 pages of findings, analyses, and detailed technical appendices. "Just as the terrible long-range weapons of the Nuclear Age made us think differently about security in the last half of the twentieth century," the report stated in its opening pages, "the electronic technology of the Information Age challenges us to invent new ways of protecting ourselves now. We must learn to negotiate a new geography, where borders are irrelevant and distances meaningless, where an enemy may be able to harm the vital systems we depend on without confronting our military power."

It went on: "Today, a computer can cause switches or valves to open and close, move funds from one account to another, or convey a military order almost as quickly over thousands of miles as it can from next door, and just as easily from a terrorist hideout as from an office cubicle or a military command center." These "cyber attacks" could be "combined with physical attacks" in an effort "to paralyze or panic large segments of society, damage our capability to respond to incidents (by disabling the 911 system or emergency communications, for example), hamper our ability to deploy conventional military forces, and otherwise limit the freedom of action of our national leadership."

The report eschewed alarmism; there was no talk here of a "cyber Pearl Harbor." Its authors allowed up front that they saw "no evidence of an *impending* cyber attack which could have a debilitating effect on the nation's critical infrastructure." Still, they added, "this is no basis for complacency," adding, "The capability to do harm—

particularly through information networks—is real; it is growing at an alarming rate; and we have little defense against it."

This was hardly the first report to issue these warnings; the conclusions reached decades earlier by Willis Ware, and adopted as policy (or attempted policy) by the Reagan administration's NSDD-145, had percolated through the small, still obscure community of technically minded officials. In 1989, eight years before General Marsh's report, the National Research Council released a study titled *Growing Vulnerability of the Public Switched Networks*, which warned that "a serious threat to communications infrastructure is developing" from "natural, accidental, capricious, or hostile agents."

Two years after that, a report by the same council, titled *Computers at Risk*, observed, "The modern thief can steal more with a computer than with a gun. Tomorrow's terrorist may be able to do more damage with a keyboard than with a bomb."

In November 1996, just eleven months before the Marsh Report came out, a Defense Science Board task force on "Information Warfare-Defense" described the "increasing dependency" on vulnerable networks as "ingredients in a recipe for a national security disaster." The report recommended more than fifty actions to be taken over the next five years, at a cost of $3 billion.

The chairman of that task force, Duane Andrews, had recently been the assistant secretary of defense for command, control, communications, and intelligence—the Pentagon's liaison with the NSA. The vice chairman was Donald Latham, who, at the ASD(C3I) twelve years earlier, had been the driving force behind Reagan's NSDD-145, the first presidential directive on computer security. In his preface, Andrews was skeptical, bordering on cynical, that the report would make a dent. "I should also point out," he wrote, "that this is the third consecutive year a DSB Summer Study or Task Force has made similar recommendations."

But unlike those studies, the Marsh Report was the work of a *presidential* commission. The commander-in-chief had ordered it into being; someone on his staff would read its report; maybe the president himself would scan the executive summary; there was, in short, a *chance* that policy would sprout from its roots.

For a while, though, there was nothing: no response from the president, not so much as a meeting or a photo op with the chairman. A few months later, Clinton briefly alluded to the report's substance in a commencement address at the Naval Academy. "In our efforts to battle terrorism and cyber attacks and biological weapons," he said, "all of us must be extremely aggressive." That was it, at least in public.

But behind the scenes, at the same time that Marsh and his commissioners were winding up their final hearings, the Pentagon and the National Security Agency were planning a top secret exercise—a simulation of a cyber attack—that would breathe life into Marsh's warnings and finally, truly, prod top officials into action.

CHAPTER 4

ELIGIBLE RECEIVER

O N June 9, 1997, twenty-five members of an NSA "Red
Team" launched an exercise called Eligible Receiver, in
which they hacked into the computer networks of the Department
of Defense, using only commercially available equipment and software. It was the first high-level exercise testing whether the U.S.
military's leaders, facilities, and global combatant commands were
prepared for a cyber attack. And the outcome was alarming.

Eligible Receiver was the brainchild of Kenneth Minihan, an Air
Force three-star general who, not quite a year and a half earlier, had
succeeded Mike McConnell as director of the NSA. Six months
before then, in August 1995, he'd been made director of the Defense Intelligence Agency, the culmination of a career in military
intel. He didn't want to move to Fort Meade, especially after such
a short spell at DIA. But the secretary of defense insisted: the NSA
directorship was more important, he said, and the nation needed
Minihan at its helm.

The secretary of defense was Bill Perry, the weapons scientist
who, back in the Carter administration, had coined and defined

"counter command-control warfare"—the predecessor to "information warfare"—and, before then, as the founding president of ESL, Inc. had built many of the devices that the NSA used in laying the grounds for that kind of warfare.

Since joining the Clinton administration, first as deputy secretary of defense, then secretary, Perry had kept an eye on the NSA, and he didn't like what he saw. The world was rapidly switching to digital and the Internet, yet the NSA was still focused too much on telephone circuits and microwave signals. McConnell had tried to make changes, but he lost focus during his Clipper Chip obsession.

"They're broken over there," Perry told Minihan. "You need to go fix things."

Minihan had a reputation as an "out-of-the-box" thinker, an eccentric. In most military circles, this wasn't seen as a good thing, but Perry thought he had the right style to shake up Fort Meade.

For a crucial sixteen-month period, from June 1993 until October 1994, Minihan had been commander at Kelly Air Force Base, sprawled out across an enclave called Security Hill on the outskirts of San Antonio, Texas, home to the Air Force Information Warfare Center. Since 1948, four years before the creation of the NSA, Kelly had been the place where, under various rubrics, the Air Force did its own code-making and code-breaking.

In the summer of 1994, President Clinton ordered his generals to start planning an invasion of Haiti. The aim, as authorized in a U.N. Security Council resolution, was to oust the dictators who had come to power through a coup d'état and to restore the democratic rule of the island-nation's elected president, Jean-Bertrand Aristide. It would be a multipronged invasion, with special operations forces pre-positioned inside the country, infantry troops swarming onto the island from several approaches, and aircraft carriers offering support offshore in the Caribbean. Minihan's task was to come up with a way for U.S. aircraft—those carrying troops and those strafing the enemy, if necessary—to fly over Haiti without being detected.

One of Minihan's junior officers in the Information Warfare Center had been a "demon-dialer" in his youth, a technical whiz kid—not unlike the Matthew Broderick character in *WarGames*—who messed with the phone company, simulating certain dial tones, so he could make long-distance calls for free. Faced with the Haiti challenge, he came to Minihan with an idea. He'd done some research: it turned out that Haiti's air-defense system was hooked up to the local telephone lines—and he knew how to make all the phones in Haiti busy at the same time. There would be no need to attack anti-aircraft batteries with bombs or missiles, which might go astray and kill civilians. All that Minihan and his crew had to do was to tie up the phone lines.

In the end, the invasion was called off. Clinton sent a delegation of eminences—Jimmy Carter, Colin Powell, and Sam Nunn—to warn the Haitian dictators of the impending invasion; the dictators fled; Aristide returned to power without a shot fired. But Minihan had woven the demon-dialer's idea into the official war plan; if the invasion had gone ahead, that was how American planes would have eluded fire.

Bill Perry was monitoring the war plan from the outset. When he learned about Minihan's idea, his eyes lit up. It resonated with his own way of thinking as a pioneer in electronic countermeasures. The Haiti phone-flooding plan was what put Minihan on Perry's radar screen as an officer to watch—and, when the right slot opened up, Perry pushed him into it.

Something else about Kelly Air Force Base caught Perry's attention. The center didn't just devise clever schemes for offensive attacks on adversaries; it also, in a separate unit, devised a clever way to detect, monitor, and neutralize information warfare attacks that adversaries might launch on America. None of the other military services, not even the Navy, had designed anything nearly so effective.

The technique was called Network Security Monitoring, and it was the invention of a computer scientist at the University of California at Davis named Todd Heberlein.

In the late 1980s, hacking emerged as a serious nuisance and an occasional menace. The first nightmare case occurred on November 2, 1988, when, over a period of fifteen hours, as many as six thousand UNIX computers—about one tenth of all the computers on the Net, including those at Wright-Patterson Air Force Base, the Army Ballistic Research Lab, and several NASA facilities—went dead and stayed dead, incurably infected from some outside source. It came to be called the "Morris Worm," named after its perpetrator, a Cornell University grad student named Robert T. Morris Jr. (To the embarrassment of Fort Meade, he turned out to be the son of Robert Morris Sr., chief scientist of the NSA Computer Security Center. It was the CSC that traced the worm to its culprit.)

Morris had meant no harm. He'd started hacking into the Net, using several university sites as a portal to hide his identity, in order to measure just how extensive the network was. (At the time, no one knew.) But he committed a serious mistake: the worm interrogated several machines repeatedly (he hadn't programmed it to stop once it received an answer), overloading and crashing the systems. In the worm's wake, many computer scientists and a few officials drew a frightening lesson: Morris had shown just how easy it was to bring the system down; had that been his intent, he could have wreaked much greater damage still.

As a result of the Morris Worm, a few mathematicians developed programs to detect intruders, but these programs were designed to protect individual computers. Todd Heberlein's innovation was designing intrusion-detection software to be installed on an open *network*, to which any number of computers might be connected. And his software worked on several levels. First, it checked for anomalous activity on the network—for instance, key words that indicated someone was making repeated attempts to log on to an account or trying out one random password after another. Such attempts drew particular notice if they entered the network with an MIT.edu address, since MIT, the Massachusetts Institute of Technology, was famous for letting anyone and everyone dial in to its terminal from anyplace on the Net and was thus a favorite point of entry for hackers. Anomalous activities would trigger an alert. At that point, the software could track data from the hacker's session, noting his IP address, how long he stayed inside the network, and how much data he was extracting or transferring to another site. (This "session date" would later be called "metadata.") After this point, if the hacker's sessions raised enough suspicion to prompt further investigation, Heberlein's software could trace their full *contents*—what the hacker was doing, reading, and sending—in real time, across the whole network that the software was monitoring.

Like many hackers and counter-hackers of the day, Heberlein had been inspired by Cliff Stoll's 1989 book, *The Cuckoo's Egg*. (A junior officer, who helped adapt Heberlein's software at the Air Force Information Warfare Center, wrote a paper called "50 Lessons from the First 50 Pages of *The Cuckoo's Egg*.") Stoll was a genial hippie and brilliant astronomer who worked at Lawrence Berkeley National Laboratory, as the computer system's administrator. One day, he discovered a seventy-five-cent error in the lab's phone bill, traced its origins out of sheer curiosity, and wound up uncovering an East German spy ring attempting to steal U.S. military secrets, using the Berkeley Lab's open site as a portal. Over the next several months, relying entirely on his wits, Stoll essentially invented the techniques of intrusion detection that came to be widely adopted over the next three decades. He attached a printer to the input lines of the lab's computer system, so that it typed a transcript of the attacker's activities. Along with a Berkeley colleague, Lloyd Bellknap, he built a "logic analyzer" and programmed it to track a specific user: when the user logged in, a device would automatically page Stoll, who would dash to the lab. The logic analyzer would also cross-correlate logs from other sites that the hacker had intruded, so Stoll could draw a full picture of what the hacker was up to.

Heberlein updated Stoll's techniques, so that he could track and repel someone hacking into not only a single modem but also a computer network.

Stoll was the inspiration for Heberlein's work in yet another sense. After Stoll gained fame for catching the East German hacker, and his book scaled the best-seller list, Lawrence Livermore National Laboratory—the more military-slanted lab, forty miles from Berkeley—exploited the headlines and requested money from the Department of Energy to create a "network security monitoring" system. Livermore won the contract, but no one there knew how to build such a system. The lab's managers reached out to Karl Levitt, a computer science professor at UC Davis. Levitt brought in his star student, Todd Heberlein.

By 1990, the Air Force Cryptology Support Center (which, a few years later, became part of the Air Force Information Warfare Center) was upgrading its intrusion-detection system. After the Morris Worm, the tech specialists started installing "host-based attack-detection" systems, the favored method of the day, which could protect a single computer; but they were quickly deemed inadequate. Some of the specialists had read about Heberlein's Network Security Monitoring software, and they commissioned him to adapt it to the center's needs.

Within two years, the cryptologists installed his software—which they renamed the Automated Security Incident Measurement, or ASIM, system—on Air Force networks. A new subdivision of the Air Force center, called the Computer Emergency Response Team, was set up to run the software, track hackers, and let higher-ups know if a serious break-in was under way. From their cubicles in San Antonio, the team could look out on Air Force networks across the nation—or that was the idea, anyway.

The program faced bureaucratic obstacles from the start. On October 7, 1992, Robert Mueller, the assistant attorney general in charge of the Justice Department's criminal division, wrote a letter, warning that network monitoring might violate federal wiretapping laws. A device that monitored a network couldn't help but pick up the Internet traffic of some innocent civilians, too. Mueller noted that the practice *might* not be illegal: the wiretapping statutes were written before the age of computer hackers and viruses; no court had yet ruled on their present-day application. But pending such a ruling, Mueller wrote, all federal agencies using these techniques should post a "banner warning," giving notice to "unauthorized intruders" that they were being monitored.

The Air Force officers in San Antonio ignored Mueller's letter: it wasn't a cease-and-desist order; and besides, warning hackers that they were being watched would destroy the whole point of a monitor.

One year later, Heberlein got a phone call from an official at the Justice Department. At first, he held his breath, wondering if at last the feds were coming to get him. To the contrary, it turned out, the department had recently installed his software, and the official had a technical question about one of its features. Justice had changed its tune, and adapted to the new world, very quickly. In a deep irony, Robert Mueller later became director of the FBI and relentlessly employed network-monitoring software to track down criminals and terrorists.

Still, back at the dawn of the new era, Mueller raised a legitimate question: Was it legal for the government to monitor a network that carried the communications not just of foreign bad guys but of ordinary Americans, too? The issue would raise its head again twenty years later, with greater passion and wider controversy, when an NSA contractor named Edward Snowden leaked a trove of ultrasecret documents detailing the agency's vast metadata program.

More daunting resistance to the network-monitoring software, in its beginnings, came from the Air Force itself. In October 1994, Minihan was transferred from Kelly to the Pentagon, where he assumed the post of Air Force intelligence chief. There, he pushed hard for wider adoption of the software, but progress was slow. Air Force computer servers had slightly more than a hundred points of entry to the Net; by the time he left the Pentagon, two years later, the computer teams back in San Antonio had received permission to monitor only twenty-six of them.

It wasn't just the monitors that Minihan had a hard time getting the top brass to accept; it was the very topic of computer security. He told three- and four-star generals about the plan to tie up the phone lines in Haiti, adding that his former teams in San Antonio were now devising similar operations against enemy *computers*. Nobody was interested. Most of the generals had risen through the ranks as pilots of fighter planes or bombers; to their way of thinking, the best way to disable a target was to drop a bomb on it. This business of hacking into computer links wasn't reliable and couldn't be measured; it reeked of "soft power." General Colin Powell may have issued a memorandum on information warfare, but they weren't buying it.

Minihan's beloved Air Force was moving too slowly, and it was way ahead of the Army and Navy in this game. His frustration had two layers: he wanted the military—all three of the main services, as well as the Pentagon's civilian leadership—to know how good his guys were at hacking the adversaries' networks; and he wanted them to know how wide open their own networks were to hacking by the same adversaries.

As the new director of the NSA, he was determined to use the job to demonstrate just how good and how bad these things were.

_____

Each year, the Pentagon's Joint Staff held an exercise called Eligible Receiver—a simulation or war game designed to highlight some threat or opportunity on the horizon. One recent exercise had focused on the danger of biological weapons. Minihan wanted the next one to test the vulnerability of the U.S. military's networks to a cyber attack. The most dramatic way to do this, he proposed, was to launch a *real* attack on those networks by a team of SIGINT specialists at the NSA.

Minihan got the idea from a military exercise, already in progress, involving the five English-speaking allies—the United States, Great Britain, Canada, Australia, and New Zealand—known in NSA circles as the "five eyes," for their formal agreement to share ultrasecret intelligence. The point of the exercise was to test new command-control equipment, some of it still in research and development. As part of this test, an eight-man crew, called the Coalition Vulnerability Assessment Team, working out of the Defense Information Systems Agency in Arlington, Virginia, would try to hack into the equipment. Minihan was told that the hackers *always* succeeded.

The assessment team's director was an American civilian named Matt Devost, twenty-three years old, a recent graduate of St. Michael's College in Burlington, Vermont, where he'd studied international relations and computer science. In his early teens, Devost had been a recreational hacker, competing with his tech friends—all of whom had watched *WarGames* several times—to see who could hack into the servers of NASA and other quasi-military agencies. Now Devost was sitting in an office with several like-minded foreigners, hacking some of the most classified systems in the world, then briefing two- and three-star generals about their exploits—all in the name of bolstering American and allied defenses.

In the most recent coalition war game, Devost's team had shut down the command-control systems of three players—Canada, Australia, and New Zealand—and taken over the American commander's personal computer, sending him fake emails and false information, thus distorting his view of the battlefield and leading him to make bad decisions, which, in a real war, could have meant defeat.

The NSA had a similar group called the Red Team. It was part of the Information Assurance Directorate (formerly called the Information Security Directorate), the defensive side of the NSA, stationed in FANEX, the building out near Friendship Airport. During its most sensitive drills, the Red Team worked out of a chamber called The Pit, which was so secret that few people at NSA knew it existed, and even they couldn't enter without first passing through two combination-locked doors. In its workaday duties, the Red Team probed for vulnerabilities in new hardware or software that had been designed for the Defense Department, sometimes for the NSA itself. These systems had to clear a high bar to be deemed secure enough for government purchase and installation. The Red Team's job was to test that bar.

Minihan's idea was to use the NSA Red Team in the same way that the five-eyes countries were using Devost's Coalition Vulnerability Assessment Team. But instead of putting it to work in a narrowly focused war game, Minihan wanted to expose the security gaps of the entire Department of Defense. He'd been trying for years to make the point to his fellow senior officers; now he wanted to hammer it home to the top officials in the Pentagon.

Bill Perry liked the idea. Still, it took Minihan a year to jump through the Pentagon bureaucracy's hoops. In particular, the general counsel needed convincing that it was legal to hack into military computers, even as part of an exercise to test their security. NSA lawyers pointed to a document called National Security Directive 42, signed by President George H. W. Bush in 1990 (as an update to

Reagan's NSDD-145), which expressly allowed such tests, as long as the secretary of defense gave written consent. Secretary Perry signed the agreement form.

The lawyers placed just one restriction on the exercise: the NSA hackers couldn't attack American networks with any of their top secret SIGINT gear; they could use only commercially available equipment and software.

On February 16, 1997, General John Shalikashvili, the chairman of the Joint Chiefs of Staff, issued Instruction 3510.01, "No-Notice Interoperability Exercise (NIEX) Program," authorizing and describing the scenario for Eligible Receiver.

The game laid out a three-phase scenario. In the first, North Korean and Iranian hackers (played by the NSA Red Team) would launch a coordinated attack on the critical infrastructures, especially the power grids and 911 emergency communication lines, of eight American cities—Los Angeles, Chicago, Detroit, Norfolk, St. Louis, Colorado Springs, Tampa, Fayetteville—and the island of Oahu, in Hawaii. (This phase was played as a tabletop game, premised on analyses of how easy it might be to disrupt the grid and overload the 911 lines.) The purpose of the attack, in the game's scenario, was to pressure American political leaders into lifting sanctions that they'd recently imposed on the two countries.

In the second part of the game, the hackers would launch a massive attack on the military's telephone, fax, and computer networks—first in U.S. Pacific Command, then in the Pentagon and other Defense Department facilities. The stated purpose was to disrupt America's command-control systems, to make it much harder for the generals to see what was going on and for the president to respond to threats with force. This phase would not be a simulation; the NSA Red Team would actually penetrate the networks.

For the three and a half months between the JCS chairman's

authorization and the actual start of the game, the NSA Red Team prepared the attack, scoping the military's networks and protocols, figuring out which computers to hack, and how, for maximum effect.

The game, its preparation and playing, was carried out in total secrecy. General Shalikashvili had ordered a "no-notice exercise," meaning that no one but those executing and monitoring the assault could know that an exercise was happening. Even inside the NSA, only the most senior officials, the Red Team itself, and the agency's lawyer—who had to approve every step the team was taking, then brief the Pentagon's general counsel and the attorney general—were let in on the secret.

At one point during the exercise, Richard Marshall, the NSA counsel, was approached by Thomas McDermott, deputy director of the agency's Information Assurance Directorate, which supervised the Red Team. McDermott informed Marshall that he was under investigation for espionage; someone on the security staff had noticed him coming in at odd hours and using the encrypted cell phone more than usual.

"*You* know why I'm here, right?" Marshall asked, a bit alarmed.

"Yes, of course," McDermott said, assuring Marshall that he'd briefed one security officer on what was happening. Even that officer was instructed not to tell his colleagues, but instead to continue going through the motions of an investigation until the game was over.

———

Eligible Receiver 97 formally got under way on Monday, June 9. Two weeks had been set aside for the exercise to unfold, with provisions for a two-week extension if necessary. But the game was over—the entire defense establishment's network was penetrated—in four *days*. The National Military Command Center—the facility that would transmit orders from the president of the United States in wartime—was hacked on the game's first day. And most of the officers manning those servers didn't even know they'd been hacked.

The NSA Red Team steered clear of only one set of targets that it otherwise might have hacked: the two dozen Air Force servers that were monitored by the computer response team analysts in San Antonio. Figuring they'd be spotted if they broke through those networks, the hackers aimed their attacks elsewhere—and intruding elsewhere turned out to be absurdly easy.

Many defense computers, it turned out, weren't protected by a password. Others were protected by the lamest passwords, like "password" or "ABCDE" or "12345." In some cases, the Red Team snipped all of an office's links except for a fax line, then flooded that line with call after call after call, shutting it down. In a few instances, NSA attachés—one inside the Pentagon, the other at a Pacific Command facility in Hawaii—went dumpster diving, riffling through trash cans and dumpsters, looking for passwords. This trick, too, bore fruit.

The team had the hardest time hacking into the server of the J-2, the Joint Staff's intelligence directorate. Finally, one of the team members simply called the J-2's office and said that he was with the Pentagon's IT department, that there were some technical problems, and that he needed to reset all the passwords. The person answering the phone gave him the existing password without hesitating. The Red Team broke in.

In most of the systems they penetrated, the Red Team players simply left a marker—the digital equivalent of "Kilroy was here." In some cases, though, they did much more: they intercepted and altered communications, sent false emails, deleted files, and reformatted hard drives. High-ranking officers who didn't know about the exercise found phone lines dead, messages sent but never received (or sent, but saying something completely different upon arrival), whole systems shut down or spitting out nonsense data. One officer who was subjected to this barrage sent his commander an email (which the Red Team intercepted), saying, "I don't trust my command-control."

This was the ultimate goal of information warfare, and Eligible Receiver revealed that it was more feasible than anyone in the world of conventional warfare had imagined.

A few weeks after it was over, an Air Force brigadier general named John "Soup" Campbell put together a postmortem briefing on the exercise. Campbell, a former F-15 fighter pilot, had been transferred to the Pentagon just as Eligible Receiver was getting under way. His new job was head of J-39, a bureau inside the operations directorate of the Joint Staff that served as a liaison between the managers of ultrasecret weapons programs and the military's combatant commanders. The Joint Staff needed someone to serve as its point man on Eligible Receiver; Campbell got the assignment.

He delivered the briefing to a small group that included senior civilian officials and the vice chiefs of the Air Force, Navy, and Marines. (The Army had decided not to participate in the exercise: a few of its officers knew they were vulnerable but didn't want to expose themselves to embarrassment; most of them dismissed the topic as a waste of time.)

Campbell's message was stark: Eligible Receiver revealed that the Defense Department was completely unprepared and defenseless for a cyber attack. The NSA Red Team had penetrated its entire network. Only a few officers had grasped that an attack was going on, and they didn't know what to do about it; no guidelines had ever been issued, no chain of command drawn up. Only one person in the entire Department of Defense, a technical officer in a Marine unit in the Pacific, responded to the attack in an effective manner: seeing that something odd was happening with the computer server, he pulled it offline at his own initiative.

After Campbell's briefing, the chief of the NSA Red Team, a Navy captain named Michael Sare, made a presentation, and, in case anyone doubted Campbell's claims, he brought along records of the intrusion—photos of password lists retrieved from dumpsters, tape recordings of phone calls in which officers blithely recited their passwords to strangers, and much more. (In the original draft of his brief, Sare noted that the team had also cracked the JCS chairman's password. Minihan, who read the draft in advance, told Sare to scratch that line. "No need to piss off a four-star," he explained.)

Everyone in the room was stunned, not least John Hamre, who had been sworn in as deputy secretary of defense at the end of July. Before then, Hamre had been the Pentagon's comptroller, where he'd gone on a warpath to slash the military budget, especially the part secretly earmarked for the NSA. Through the 1980s, as a staffer for the Congressional Budget Office and the Senate Armed Services Committee, Hamre had grown to distrust the NSA: it was a dodgy outfit, way too covert, floating in the gray area between "military" and "intelligence" and evading the strictures on both. Hamre didn't know anything about information warfare, and he didn't care.

A few weeks before Eligible Receiver, as Hamre prepared for his promotion, Minihan had briefed him on the threats and opportunities of information warfare and on the need for a larger budget to exploit them. Hamre, numbed by the technical detail, had sighed and said, "Ken, you're giving me a headache."

But now, listening to Campbell and Sare run down the results of Eligible Receiver, Hamre underwent a conversion, seized with a sense of urgency. Looking around the room of generals and colonels, he asked who was in charge of fixing this problem.

They all looked back at him. No one knew the answer. No one was in charge.

Around the same time, Ken Minihan delivered his own briefing on Eligible Receiver to the Marsh Commission. The panel, by now, had delved deeply into the fragile state of America's critical infrastructure. But the scenarios they'd studied were hypothetical and dealt with the vulnerability of *civilian* sectors; no one had ever launched an actual cyber attack, and most of the commissioners had assumed that the military's networks were secure. Minihan's briefing crushed their illusions on both counts: an NSA Red Team had launched an actual attack, and its effects were devastating.

Minihan did not reveal one episode of Eligible Receiver, an incident that only a few officials knew about: when the Red Team members were hacking into the networks as part of the exercise, they came across some strangers—traceable to French Internet addresses—hacking into the network for real. In other words, foreign spies were already penetrating vital and vulnerable networks; the threat wasn't hypothetical.

Even without this tidbit, the commissioners were stunned. Marsh asked what could be done to fix the problem. Minihan replied, "Change the law, give me the power, I'll protect the nation."

No one quite knew what he meant. Or, if he meant what they thought he meant, nobody took it seriously: nobody was going to revive Reagan's NSDD-145 or anything like it.

On October 13, the Marsh Commission published its report. Titled *Critical Foundations*, it only briefly alluded to Eligible Receiver. Its recommendations focused mainly on the need for the government and private industry to share information and solve problems jointly. It said nothing about giving the NSA more money or power.

Four months later, another attack on defense networks occurred—something that looked like Eligible Receiver, but coming from real, unknown hackers in the real, outside world.

CHAPTER 5

SOLAR SUNRISE, MOONLIGHT MAZE

O N February 3, 1998, the network monitors at the Air Force Information Warfare Center in San Antonio sounded the alarm: someone was hacking into a National Guard computer at Andrews Air Force Base on the outskirts of Washington, D.C.

Within twenty-four hours, the center's Computer Emergency Response Team, probing the networks more deeply, detected intrusions at three other bases. Tracing the hacker's moves, the team found that he'd broken into the network through an MIT computer server. Once inside the military sites, he installed a "packet sniffer," which collected the directories of usernames and passwords, allowing him to roam the entire network. He then created a back door, which let him enter and exit the site at will, downloading, erasing, or distorting whatever data he wished.

The hacker was able to do all this because of a well-known vulnerability in a widely used UNIX operating system. The computer specialists in San Antonio had been warning senior officers of this vulnerability—Ken Minihan had personally repeated these warnings to generals in the Pentagon—but no one paid attention.

When President Clinton signed the executive order on "Critical Infrastructure Protection," back in July 1996, one consequence was the formation of the Marsh Commission, but another—less noticed at the time—was the creation of the Infrastructure Protection Task Force inside the Justice Department, to include personnel from the FBI, the Pentagon (the Joint Staff and the Defense Information Systems Agency), and, of course, the National Security Agency.

By February 6, three days after the intrusion at Andrews Air Force Base was spotted, this task force was on the case, with computer forensics handled by analysts at NSA, DISA, and a unit in the Joint Staff called the Information Operations Response Cell, which had been set up just a week earlier as a result of Eligible Receiver. They found that the hacker had exploited a specific vulnerability in the UNIX systems, known as Sun Solaris 2.4 and 2.6. And so, the task force code-named its investigation Solar Sunrise.

John Hamre, the deputy secretary of defense who'd seen the Eligible Receiver exercise eight months earlier as the wake-up call to a new kind of threat, now saw Solar Sunrise as the threat's fulfillment. Briefing President Clinton on the intrusion, Hamre warned that Solar Sunrise might be "the first shots of a genuine cyber war," adding that they may have been fired by Iraq.

It wasn't a half-baked suspicion. Saddam Hussein had recently expelled United Nations inspectors who'd been in Iraq for six years to ensure his compliance with the peace terms that ended Operation Desert Storm—especially the clause that barred him from developing weapons of mass destruction. Many feared that Saddam's ouster of the inspectors was the prelude to resuming his WMD program. Clinton had ordered his generals to plan for military action; a second aircraft carrier was steaming to the Persian Gulf; American troops were prepared for possible deployment.

So when the Solar Sunrise hack expanded to more than a dozen military bases, it struck some, especially inside the Joint Staff, as a pattern. The targets included bases in Charleston, Norfolk, Dover, and Hawaii—key deployment centers for U.S. armed forces. Only unclassified servers were hacked, but some of the military's vital support elements—transportation, logistics, medical teams, and the defense finance system—ran on unclassified networks. If the hacker corrupted or shut down these networks, he could impede, maybe block, an American military response.

Then came another unsettling report: NSA and DISA forensics analysts traced the hacker's path to an address on Emirnet, an Internet service provider in the United Arab Emirates—lending weight to fears that Saddam, or some proxy in the region, might be behind the attacks.

The FBI's national intelligence director sent a cable to all his field agents, citing "concern that the intrusions may be related to current U.S. military actions in the Persian Gulf." At Fort Meade, Ken Minihan came down firmer still, telling aides that the hacker seemed to be "a Middle Eastern entity."

Some were skeptical. Neal Pollard, a young DISA consultant who'd studied cryptology and international relations in college, was planning a follow-on exercise to Eligible Receiver when Solar Sunrise, a real attack, took everyone by surprise. As the intrusions spread, Pollard downloaded the logs, drafted briefings, tried to figure out the hacker's intentions—and, the more he examined the data, the more he doubted that this was the work of serious bad guys.

In the exercise that he'd been planning, a Red Team was going to penetrate an unclassified military network, find a way in to its classified network (which, Pollard knew from advance probing, wasn't very secure), hop on it, and crash it. By contrast, the Solar Sunrise hacker wasn't doing anything remotely as elaborate: this guy would poke around briefly in one unclassified system after another, then get out, leaving behind no malware, no back door, nothing. And while some of the servers he attacked were precisely where a hacker would go to undermine the network of a military on the verge of deployment, most of the targets seemed selected at random, bearing no significance whatever.

Still, an international crisis was brewing, war might be in the offing; so worst-case assumptions came naturally. Whatever the hacker's identity or motive, his work was throwing commanders off balance. They remembered Eligible Receiver, when they didn't know they'd been hacked; the NSA Red Team had fed some of them false messages, which they'd assumed were real. This time around, they knew they were being hacked, and it wasn't a game. They didn't *detect* any damage, but how could they be *sure*? When they read a message or looked at a screen, could they trust—*should* they trust—what they were seeing?

This was the desired effect of what Perry had called counter command-control warfare: just knowing that you'd been hacked, regardless of its tangible effects, was disorienting, disrupting.

Meanwhile, the Justice Department task force was tracking the hacker twenty-four hours a day. It was a laborious process. The hacker was hopping from one server to another to obscure his identity and origins; the NSA had to report all these hops to the FBI, which took a day or so to investigate each report. At this point, no one knew whether Emirnet, the Internet service provider in the United Arab Emirates, was the source of the attacks or simply one of several landing points along the hacker's hops.

Some analysts in the Joint Staff's new Information Operations Response Cell noticed one pattern in the intrusions: they all took place between six and eleven o'clock at night, East Coast time. The analysts calculated what time it might be where the hacker was working: he might, it turned out, be on the overnight shift in Baghdad or Moscow, or maybe the early morning shift in Beijing.

One possibility they didn't bother to consider: it was also after-school time in California.

By February 10, after just four days of sleuthing, the task force found the culprits. They weren't Iraqis or "Middle Eastern entities" of any tribe or nationality. They were a pair of sixteen-year-old boys in the San Francisco suburbs, malicious descendants of the Matthew Broderick character in *WarGames,* hacking the Net under the usernames Makaveli and Stimpy, who'd been competing with their friends to hack into the Pentagon the fastest.

In one day's time, FBI agents obtained authority from a judge to run a wiretap. They took the warrant to Sonic.net, the service provider the boys were using, and started tracking every keystroke the boys typed, from the instant they logged on through the phone line of Stimpy's parents. Physical surveillance teams confirmed that the boys were in the house—eyewitness evidence of their involvement, in case a defense lawyer later claimed that the boys were blameless and that someone else must have hacked into their server.

Through the wiretap, the agents learned that the boys were getting help from an eighteen-year-old Israeli, an already notorious hacker named Ehud Tenenbaum, who called himself The Analyzer. All three teenagers were brazen—and stupid. The Analyzer was so confident in his prowess that, during an interview with an online forum called AntiOnline (which the FBI was monitoring), he gave a live demonstration of hacking into a military network. He also announced that he was training the two boys in California because he was "going to retire" and needed successors. Makaveli gave an interview, too, explaining his own motive. "It's power, dude," he typed out. "You know, power."

The Justice Department task force was set to let the boys hang themselves a bit longer, but on February 25, John Hamre spoke to reporters at a press breakfast in Washington, D.C. Still frustrated with the military's inaction on the broader cyber threat, he outlined

the basic facts of Solar Sunrise (which, until then, had been kept secret), calling it "the most organized and systematic attack" on American defense systems to date. And he disclosed that the suspects were
two teenagers in Northern California.

At that point, the FBI had to scramble before the boys heard about
Hamre's remarks and erased their files. Agents quickly obtained a
search warrant and entered Stimpy's house. There he was, in his
bedroom, sitting at a computer, surrounded by empty Pepsi cans and
half-eaten cheeseburgers. The agents arrested the boys while carting
off the computer and several floppy disks.

Stimpy and Makaveli (whose real names were kept under seal,
since they were juveniles) were sentenced to three years probation
and a hundred hours of community service; they were also barred
from going on the Internet without adult supervision. Israeli police arrested Tenenbaum and four of his apprentices, who were all
twenty years old; he served eight months in prison, after which he
started an information security firm, then moved to Canada, where
he was arrested for hacking into financial sites and stealing credit
card numbers.

At first, some U.S. officials were relieved that the Solar Sunrise
hackers turned out to be just a couple of kids—or, as one FBI official
put it in a memo, "not more than the typical hack du jour." But most
officials took that as the opposite of reassurance: if a couple of kids
could pull this off, what could a well-funded, hostile nation-state do?

They were about to find out.

———

In early March, just as officials at NSA, DISA, and the Joint Staff's
Information Operations Response Cell were closing their case
files on Solar Sunrise and going back to their workaday tasks,
word came through that someone had hacked into the computers at Wright-Patterson Air Force Base, in Ohio, and was pilfering

files—unclassified but sensitive—on cockpit design and microchip schematics.

Over the next few months, the hacker fanned out to other military facilities. No one knew his location (the hopping from one site to another was prodigious, swift, and global); his searches bore no clear pattern (except that they involved high-profile military R&D projects). The operation was a sequel of sorts to Solar Sunrise, though more elaborate and puzzling; so, just as night follows day, the task force called it Moonlight Maze.

Like the Solar Sunrise gang, this hacker would log in to the computers of university research labs to gain access to military sites and networks. But in other ways, he didn't seem at all like some mischievous kid on a cyber joyride. He didn't dart in and out of a site; he was persistent; he was looking for specific information, he seemed to know where to find it, and, if his first path was blocked, he stayed inside the network, prowling for other approaches.

He was also remarkably sophisticated, employing techniques that impressed even the NSA teams that were following his moves. He would log on to a site, using a stolen username and password; when he left, he would rewrite the log so that no one would know he'd ever been there. Finding the hacker was touch-and-go: the analysts would have to catch him in the act and track his moves in real time; even then, since he erased the logs when exiting, the on-screen evidence would vanish after the fact. It took a while to convince some higher-ups that there had *been* an intrusion.

A year earlier, the analysts probably wouldn't have detected a hacker at all, unless by pure chance. About a quarter of the servers in the Air Force were wired to the network security monitors in San Antonio; but most of the Army, Navy, and civilian leaders in the Pentagon would have had no way of knowing whether an intruder was present, much less what he was doing or where he was from.

That all changed with the one-two-three punch of Eligible Receiver, the Marsh Commission Report, and Solar Sunrise—which, over a mere eight-month span, from June 1997 to February 1998, convinced high-level officials, even those who had never thought about the issue, that America was vulnerable to a cyber attack and that this condition endangered not only society's critical infrastructure but also the military's ability to act in a crisis.

Right after Eligible Receiver, John Hamre called a meeting of senior civilians and officers in the Pentagon to ask what could be done. One solution, a fairly easy gap-filler, was to authorize an emergency purchase of devices known as intrusion-detection systems or IDS—a company in Atlanta, Georgia, called Internet Security Systems, could churn them out in quantity—and to install them on more than a hundred Defense Department computers. As a result, when Solar Sunrise and Moonlight Maze erupted, far more Pentagon personnel saw what was happening, far more quickly, than they otherwise would have.

Not everyone got the message. After Eligible Receiver, Matt Devost, who'd led the aggressor team in war games testing the vulnerability of American and allied command-control systems, was sent to Hawaii to clean up the networks at U.S. Pacific Command headquarters, which the NSA Red Team had devastated. Devost found gaps and sloppiness everywhere. In many cases, software vendors had long ago issued warnings about the vulnerabilities along with patches to fix them; the user had simply to push a button, but no one at PacCom had done even that. Devost lectured the admirals, all of them more than twice his age. This wasn't rocket science, he said. Just put someone in charge and order him to install the repairs. When Solar Sunrise erupted, Devost was working computer forensics at the Defense Information Systems Agency. He came across PacCom's logs and saw that they still hadn't fixed their problems: despite his strenuous efforts, nothing had changed. (He decided at that point to quit government and do computer-attack simulations in the private sector.)

Even some of the officers who'd made the changes, and installed the devices, didn't understand what they were doing. Six months after the order went out to put intrusion-detection systems on Defense Department computers (still a few weeks before Solar Sunrise), Hamre called a meeting to see how the devices were working.

An Army one-star general furrowed his brow and grumbled that he didn't know about these IDS things: ever since he'd put them on his computers, they were getting attacked every day.

The others at the table suppressed their laughter. The general didn't realize that his computers might have been getting hacked every day for months, maybe years; all the IDS had done was to let him know it.

Early on in Solar Sunrise, Hamre called another meeting, imbued with the same sweat of urgency as the one he'd called in the wake of Eligible Receiver, and asked the officers around him the same question he'd asked before: "Who's in charge?"

They all looked down at their shoes or their notepads, because, in fact, nothing had changed; no one was still in charge. The IDS devices may have been in place, but no one had issued protocols on what to do if the alarm went off or how to distinguish an annoying prank from a serious attack.

Finally, Brigadier General John "Soup" Campbell, the commander of the secret J-39 unit, who'd been the Joint Staff's point man on Eligible Receiver, raised his hand. "I'm in charge," he said, though he had no idea what that might mean.

By the time Moonlight Maze started wreaking havoc, Campbell was drawing up plans for a new office called Joint Task Force-Computer Network Defense—or JTF-CND. Orders to create the task force had been signed July 23, and it had commenced operations on December 10. It was staffed with just twenty-three officers, a mix of computer specialists and conventional weapons operators who had to take a crash course on the subject, all crammed into a trailer behind DISA headquarters in the Virginia suburbs, not far from the Pentagon. It was an absurdly modest effort for an outfit that, according to its charter, would be "responsible for coordinating and directing the defense of DoD computer systems and computer networks," including "the coordination of DoD defensive actions" with other "government agencies and appropriate private organizations."

Campbell's first steps would later seem elementary, but no one had ever taken them—few had thought of them—on such a large scale. He set up a 24/7 watch center, established protocols for alerting higher officials and combatant commands of a cyber intrusion, and—the very first step—sent out a communiqué, on his own authority, advising all Defense Department officials to change their computer passwords.

By that point, Moonlight Maze had been going on for several months, and the intruder's intentions and origins were still puzzling. Most of the intrusions, the ones that were noticed, took place in the same nine-hour span. Just as they'd done during Solar Sunrise, some intelligence analysts in the Pentagon and the FBI looked at a time zone map, did the math, and guessed that the attacker must be in Moscow. Others, in the NSA, noted that Tehran was in a nearby time zone and made a case for Iran as the hacker's home.

Meanwhile, the FBI was probing all leads. The hacker had hopped through the computers of more than a dozen universities—the University of Cincinnati, Harvard, Bryn Mawr, Duke, Pittsburgh, Auburn, among others—and the bureau sent agents to interview students, tech workers, and faculty on each campus. A few intriguing suspects were tagged here and there—an IT aide who answered questions nervously, a student with a Ukrainian boyfriend—but none of the leads panned out. The colleges weren't the source of the hack; like the Lawrence Berkeley computer center in Cliff Stoll's *The*

Cuckoo's Egg, they were merely convenient transit points from one target site to another.

Finally, three breakthroughs occurred independently. One was inspired by Stoll's book. Stoll had captured the East German hacker a dozen years earlier by creating a "honey pot"—a set of phony files, replete with directories, documents, usernames, and passwords (all of Stoll's invention), seemingly related to the American missile-defense program, a subject of particular interest to the hacker. Once lured to the pot, he stayed in place long enough for the authorities to trace his movements and track him down. The interagency intelligence group in charge of solving Moonlight Maze—mainly NSA analysts working under CIA auspices—decided to do what Stoll had done: they created a honey pot, in this case a phony website of an American stealth aircraft program, which they figured might lure their hacker. (Everyone in the cyber field was enamored of *The Cuckoo's Egg*; when Stoll, a long-haired Berkeley hippie, came to give a speech at NSA headquarters not long after his book was published, he received a hero's welcome.) Just as in Stoll's scheme, the hacker took the bait.

But with their special access to exotic tools, the NSA analysts took Stoll's trick a step further. When the hacker left the site, he unwittingly took with him a digital beacon—a few lines of code, attached to the data packet, which sent back a signal that the analysts could follow as it piggybacked through cyberspace. The beacon was an experimental prototype; sometimes it worked, sometimes it didn't. But it worked well enough for them to trace the hacker to an IP address of the Russian Academy of Sciences, in Moscow.

Some intelligence analysts, including at NSA, remained skeptical, arguing that the Moscow address was just another hopping point along the way to the hacker's real home in Iran.

Then came the second breakthrough. While Soup Campbell was setting up Joint Task Force-Computer Network Defense, he hired a

naval intelligence officer named Robert Gourley to be its intel chief.
Gourley was a hard-driving analyst with a background in computer
science. In the waning days of the Cold War, he'd worked in a unit
that fused intelligence and operations to track, and aggressively chase,
Russian submarines. He'd learned of this fusion approach, five years
earlier, at an officers' midcareer course taught by Bill Studeman and
Rich Haver—the intelligence veterans who, a decade earlier, under
the tutelage of Admiral Bobby Ray Inman, had pushed for the adoption of counter command-control warfare.

Shortly before joining Campbell's task force, Gourley attended
another conference, this one lasting just a day, on Navy operations
and intelligence. Studeman and Haver happened to be among the
lecturers. Gourley went up to them afterward to renew his ac-
quaintance. A few weeks later, ensconced in his task force office, he
phoned Haver on a secure line, laid out the Moonlight Maze problem, as well as the debate over the intruder's identity, and asked if he
had advice on how to resolve it.

Haver recalled that, during the Cold War, the KGB or GRU, the
Soviet military's spy agency, often dispatched scientists to international conferences to collect papers on topics of interest. So Gourley assembled a small team of analysts from the various intelligence
agencies and scoured the logs of Moonlight Maze to see what topics interested this hacker. The swath, it turned out, covered a bi-
zarrely wide range: not just aeronautics (the topic of his first search,
at Wright-Patterson) but also hydrodynamics, oceanography, the al-
timeter data of geophysical satellites, and a lot of technology related
to surveillance imagery. Gourley's team then scanned databanks of
recent scientific conferences. The matchup was at least intriguing:
Russian scientists had attended conferences on every topic that at-
tracted the hacker.

That, plus the evidence from the honey pot and the absence of
signs pointing to Iran or any other Middle Eastern source, led Gour-

ley to conclude that the culprit was Russia. It was a striking charge: a *nation-state* was hacking American military networks—and not just any nation-state, but America's former enemy and now, supposedly, post–Cold War partner.

Gourley brought his finding to Campbell, who was shocked. "Are you saying that we're under attack?" he asked. "Should we declare war?"

"No, no," Gourley replied. This was an intelligence *assessment*, though he added that he had "high confidence" in its accuracy.

The third intelligence breakthrough was the firmest but also the newest, the one that relied on methods unique to the cyber age and thus mastered by only a few fledgling specialists. Kevin Mandia was part of a small cyber crime team at the Air Force Office of Special Investigations. He'd visited the Air Force Information Warfare Center in San Antonio several times and had kept up with its network security monitoring system. When Moonlight Maze got started, Mandia, by now a private contractor, was sent to the FBI task force to review the hacker's logs. The hacker was using an obfuscated code; Mandia and his team wrote new software to decrypt the commands—and it turned out they'd been typed in Cyrillic. Mandia concluded that the hacker was Russian.*

For the first several months of Moonlight Maze, the American intelligence agencies stopped short of making any statement, even informally, about the hacker's origins. But the convergence of the Stoll-inspired honey pot, Bob Gourley's analysis, and Kevin Mandia's decryption—the fact that such disparate methods sired the same conclusion—changed the picture. It was also clear by now that the Moonlight Maze hackers, whoever they were, had pulled in quite

---

*In 2006, Mandia would form a company called Mandiant, which would emerge as one of the leading cyber security incident consultants, rising to prominence in 2011 as the firm that identified a special unit of the Chinese army as the hacker behind hundreds of cyber attacks against Western corporations.

a haul: 5.5 gigabytes of data, the equivalent of nearly three million sheets of paper. None of it was classified, but quite a lot of it was sensitive—and might add up to classified information if a smart analyst pieced it all together.

For nearly a year, an FBI-led task force—the same interagency task force that investigated Solar Sunrise—had coordinated the interagency probe, sharing all intelligence and briefing the White House. In February, John Hamre testified on the matter in closed hearings. Days later, the news leaked to the press, including the finding that the hackers were Russian.

At that point, some members of the task force, especially those from the FBI, proposed sending a delegation to Moscow and confronting Russian officials head-on. It might turn out that they had nothing to do with the hacking (Hamre had testified that it was unclear whether the hackers were working in the government), in which case the Kremlin and the security ministries would want to know about the renegade in their midst. Or maybe the Russian government *was* involved, in which case that would be worth knowing, too.

Task force members from the Pentagon and NSA were leery about going public. Maybe the Russians hadn't read the news stories, or maybe they had but dismissed the reports as untrue; in other words, maybe the Russians still didn't know we were on to them, that we were hacking their hacker. Meanwhile, we were learning things about their interests and operational style; an official confrontation could blow the operation.

In the end, the White House approved the FBI's request to send a delegation. The task force then spent weeks discussing what evidence to let the Russians see and what evidence to withhold. In any case, it would be presented to the Russians in the same terms as the FBI officially approached it—not as a matter of national security or diplomacy, but rather as a *criminal* investigation, in which the United States was seeking assistance from the Russian Federation.

The delegation, formally called the Moonlight Maze Coordination Group, consisted of four FBI officials—a field agent from the Baltimore office, two linguists from San Francisco, and a supervisor from headquarters—as well as a NASA scientist and two officers from the Air Force Office of Special Investigations, who had examined the hacker's logs with Kevin Mandia. They flew to Moscow on April 2, bringing along the files from five of the cyber intrusions, with plans to stay for eight days.

This was the era of warm relations between Bill Clinton and Russia's reform president, Boris Yeltsin, so the group was received in a spirit of celebration, its first day in Moscow filled with toasts, vodka, caviar, and good cheer. They spent the second day at the headquarters of the Russian defense ministry in a solid working session. The Russian general who served as the group's liaison was particularly cooperative. He brought out the logs on the files that the Americans had brought with them. This was confirmation: the Russian government had been the hacker, working through servers of the academy of sciences. The general was embarrassed, putting blame on "those motherfuckers in intelligence."

As a test, to see whether this might be a setup, one of the Air Force investigators on the trip mentioned a sixth intrusion, one whose files the group hadn't brought with them. The general brought out those logs, too. This is criminal activity, he bellowed to his new American friends. We don't tolerate this.

The Americans were pleased. This was working out extraordinarily well; maybe the whole business could be resolved through quiet diplomacy and a new spirit of cooperation.

On the third day, things took a shaky turn. Suddenly, the group's escorts announced that it would be a day of sightseeing. So was the fourth day. On the fifth day, no events were scheduled at all. The Americans politely protested, to no avail. They never again stepped foot inside the Russian defense ministry. They never again heard from the helpful general.

As they prepared to head back to the States, on April 10, a Russian officer assured them that his colleagues had launched a vigorous investigation and would soon send the embassy a letter outlining their findings.

For the next few weeks, the legal attaché in the American embassy phoned the Russian defense ministry almost every day, asking if the letter had been written. He was politely asked to be patient. No letter ever arrived. And the helpful general seemed to have vanished.

Back in Washington, a task force member cautioned against drawing sour conclusions. Maybe, he said, the general was just sick.

Some members from the Pentagon and the intelligence agencies, who'd warned against the trip, rolled their eyes. "Yeah," Bob Gourley scoffed, "maybe he has a case of lead poisoning."

The emerging consensus was that the general hadn't known about the hacking operation, that he'd genuinely believed some recalcitrant agents in military intelligence were engaged in skullduggery—until his superiors excoriated him, possibly fired him or worse, for sharing secrets with the Americans.

One good thing came out of the trip: the hacking did seem to stop.

Then, two months later, Soup Campbell's Joint Task Force-Computer Network Defense detected another round of hacking into sensitive military servers—these intrusions bearing a slightly different signature, layered with codes that were harder to break.

The cat-and-mouse game was back on. And it was a game where both sides, and soon other nations, played cat *and* mouse. To an extent known by only a few American officers, still fewer political higher-ups, and no doubt some Russian spies, too, the American cyber warriors were playing offense as well as defense—and had been for a long while.

CHAPTER 6

THE COORDINATOR MEETS MUDGE

I N October 1997, a few months before Solar Sunrise, when the Marsh Commission released its report on the nation's critical infrastructure, few officials were more stunned by its findings than a White House aide named Richard Alan Clarke.

As the counterterrorism adviser to President Clinton, Clarke had been in on the high-level discussions after the Oklahoma City bombing and the subsequent drafting of PDD-39, Clinton's directive on counterterrorism, which eventually led to the formation of the Marsh Commission. After that, Clarke returned to his usual routines, which mainly involved tracking down a Saudi jihadist named Osama bin Laden.

Then the Marsh Report came out, and most of it dealt with *cyber* security. It was a topic Clarke had barely heard of. Still, it wasn't his topic. Rand Beers, a good friend and Clinton's intelligence adviser, had been the point man on the commission and, presumably, would deal with the report, as well. But soon after its release, Beers announced that he was moving over to the State Department; he and Sandy Berger, Clinton's national security adviser, had discussed who should replace him on the cyber beat, and they settled on Clarke.

Clarke resisted; he was busy enough on the bin Laden trail. Then again, he had been the White House point man on the Eligible Receiver exercise; Ken Minihan, the NSA director who'd conceived it, had briefed him thoroughly on its results and implications; cyber security might turn out to be interesting. But Clarke knew little about computers or the Internet. So he gathered a few of his staff and took them on a road trip.

Shortly after the holidays, they flew to the West Coast and visited the top executives of the major computer and software firms. What struck Clarke most was that the heads of Microsoft knew all about operating systems, those at Cisco knew all about routers, those at Intel knew all about chips—but none of them seemed to know much about the gadgets made by the others or the vulnerabilities at the seams in between.

Back in Washington, he asked Minihan for a tour of the NSA. Clarke had been a player in national security policy for more than a decade, since the Reagan administration, but for most of that time, he'd been involved in Soviet-American arms-control talks and Middle East crises: the high-profile issues. He'd never had reason to visit, or think much about, Fort Meade. Minihan told his aides to give Clarke the full dog-and-pony show.

Part of the tour was demonstrating how easily the SIGINT teams could penetrate any foreign network they set their eyes on. None of it reassured Clarke; he came away more shaken than before, for the same reason as many officials who'd witnessed similar displays through the years. If we can do this to other countries, he realized, they'll soon be able to do the same thing to us—and that meant we were screwed, because *nothing* on the Internet could be secured, and, as the Marsh Report laid out in great detail, everything in America was going up on the Net.

Clarke wanted to know just how vulnerable America's networks were right now, and he figured the best way to find out was to talk with some hackers. He didn't want to deal with criminals, though, so he called a friend at the FBI and asked if he knew any good-guy hackers. (At this point, Clarke didn't know if such creatures existed.) At first, the agent was reluctant to share sources, but finally he put Clarke in touch with "our Boston group," as he put it—a team of eccentric computer geniuses who occasionally helped out with law-enforcement investigations and who called themselves "The L0pht" (pronounced "loft").

The L0pht's front man—who went by the handle "Mudge"—would meet Clarke at John Harvard's Brewery, near Harvard Square, in Cambridge, on a certain day at seven p.m. Clarke flew to Boston on the designated day, took a cab to the bar, and settled in at seven on the dot. He waited an hour for someone to approach him; no one did; so he got up to leave, when the man quietly sitting next to him touched his elbow and said, "Hi, I'm Mudge."

Clarke looked over. The man, who seemed about thirty, wore jeans, a T-shirt, one earring, a goatee, and long golden hair ("like Jesus," he would later recall).

"How long have you been sitting there?" Clarke asked.

"About an hour," Mudge replied. He'd been there the whole time.

They chatted casually about the L0pht for a half hour or so, at which point Mudge asked Clarke if he'd like to meet the rest of the group. Sure, Clarke replied. They're right over there, Mudge said, pointing to a large table in the corner where six guys were sitting, all in their twenties or early thirties, some as unruly as Mudge, others clean-cut.

Mudge introduced them by their tag names: Brian Oblivion, Kingpin, John Tan, Space Rogue, Weld Pond, and Stefan von Neumann.

After some more small talk, Mudge asked Clarke if he'd like to see the L0pht. Of course, he replied. So they took a ten-minute drive to what looked like a deserted warehouse in Watertown, near the

Charles River. They went inside, walked upstairs to the second floor, unlocked another door, and turned on the lights, which revealed a high-tech laboratory, crammed with dozens of mainframe computers, desktops, laptops, modems, and a few oscilloscopes, much of it wired—as Mudge pointed out, when they went back outside—to an array of antennas and dishes on the roof.

Clarke asked how they could afford all this equipment. Mudge said it didn't cost much. They knew when the big computer companies threw out hardware (a few of them worked for these companies under their real names); they'd go to the dumpster that day, retrieve the gear, and refurbish it.

The collective had started, Clarke learned, in the early 1990s, mainly as a place where its members could store their computers and play online games. In 1994, they made a business of it, testing the big tech firms' new software programs and publishing a bulletin that detailed the security gaps. They also designed, and sold for cheap, their own software, including L0phtCrack, a popular program that let buyers crack most passwords stored on Microsoft Windows. Some executives complained, but others were thankful: *someone* was going to find those flaws; at least the L0pht was doing it in the open, so the companies could fix them. The NSA, CIA, FBI, and the Air Force Information Warfare Center were also intrigued by this guerrilla operation; some of their agents and officers started talking with Mudge, who'd emerged as the group's spokesman, and even invited him to give talks at high-level security sessions.

Not that the intelligence agencies needed Mudge to tell them about holes in commercial software. The cryptologists in the NSA Information Assurance Directorate spent much of their time probing for these holes; they'd found fifteen hundred points of vulnerability in Microsoft's first Windows system. And, by an agreement much welcomed by the software industry at the time, they routinely told the firms about their findings—most of the findings, anyway: they always left a few holes for the agency's SIGINT teams to exploit, since the foreign governments that they spied on had bought this software, too. (Usually, the Silicon Valley firms were complicit in leaving back doors open.) Still, the NSA and the other agencies were interested in how the likes of Mudge were tackling the problem; it gave them insights into ways that other, more malicious, perhaps foreign hackers might be operating, ways that their own security specialists might not have considered.

For his part, Mudge was always happy to give them advice and never charged a fee. He figured that, any day now, the feds could come knocking at the warehouse door—some of the L0pht gang's projects were of dubious legality—and it would be useful to summon, as character witnesses, the directors of the nation's intelligence and law enforcement agencies.

For the next few hours on that winter night in Watertown, the L0pht gang held Clarke's rapt attention, telling him all the things they could do, if they wanted. They could break the passwords stored on any operating system, not just Microsoft Windows. They could decrypt any satellite communications. They had devised software (not yet for sale or distribution) that could hack into someone's computer and control it remotely, spying on the user's every keystroke, changing his files, tossing him off the Internet or whisking him away to a site of their choosing. They had special machines that let them reverse-engineer any microchip by de-capping the chip and extracting the silicon dye. In hushed tones, they told him about a recent discovery, involving the vulnerability of the Border Gateway Protocol, a sort of supra-router for all online traffic, which would let them—or some other skilled hackers—shut down the entire Internet in a half hour.

Clarke didn't know whether to believe everything they said, but he was awed and shaken. Everyone who'd briefed him, during his crash course on the workings and pitfalls of the Internet, had implied or stated outright that only nation-states possessed the resources to do just half the things that Mudge and his chums were saying—and, in some cases, demonstrating—that they could do from this hole in the wall with little money and, as far as he could tell, no outside support. In short, the official threat model seemed to have it all wrong.

And Clarke, the president's special adviser on counterterrorism, realized that this cyber thing was more than an engrossing diversion; it fit into his bailiwick precisely. If Mudge and his gang used their talents to disrupt American society and security, exploiting the critical vulnerabilities that the Marsh Report had outlined, they would be tagged as terrorists—*cyber terrorists*. Here, then, was another threat for Clarke to worry about—and to add to his thickening portfolio.

It was two a.m., after a few more drinks, when they decided to call it a night. Clarke asked them if they'd like to come down to Washington for a private tour of the White House, and he offered to pay their way.

Mudge and the others were startled. "Hackers"—which is what they were—was still a nasty term in most official corridors. It was one thing for some spook in a three-letter agency to invite them to brief a roomful of other spooks on a hush-hush basis—quite another to be invited to the White House by a special adviser to the president of the United States.

A month later, they came down, not only to see the West Wing after hours but also to testify before Congress. The Senate Governmental Affairs Committee happened to be holding hearings on cyber security. Through his Hill contacts, Clarke got the L0pht members—all seven of them, together, using their pseudonyms—placed on the witness list.

Clarke had a few more conversations with Mudge during this period. His real name, it turned out, was Peiter Zatko. He'd been a hacker since his early teen years. He hated the movie *WarGames* because it encouraged too many other people his age, but nowhere near his IQ, to join the field. He'd graduated not from someplace like MIT, as Clarke suspected, but from the Berklee College of Music, as a guitar major, at the top of his class. By day, Zatko was working as a computer security specialist at BNN, a Cambridge-based firm, though his looming public profile accelerated his plans to quit and turn the L0pht into a full-time commercial enterprise.

He and the other L0pht denizens made their Capitol Hill debut on May 19, 1998. Only three senators attended the hearing—the chairman Fred Thompson, John Glenn, and Joe Lieberman—but they treated the bizarre witnesses with respect, hailing them as patriots, Lieberman likening them to Paul Revere, alerting the citizenry to danger in the digital age.

Three days after Mudge's testimony, Clinton signed a Presidential Decision Directive, PDD-63, titled "Critical Infrastructure Protection," reprising the Marsh Commission's findings—the nation's growing dependence on computer networks, the vulnerability of those networks to attack—and outlining ways to mitigate the problem.

A special panel of the NSC, headed by Rand Beers, had cut-and-pasted early drafts of the directive. Then, at one of the meetings, Beers informed the group that he was moving to the State Department and that Dick Clarke—who, for the first time, was seated next to him—would take his place on the project.

Several officials on the panel raised their eyebrows. Clarke was a brash, haughty figure, a spitball player of bureaucratic politics, admired by some as a can-do operator, despised by others as a power-grabbing manipulator. John Hamre, the deputy secretary of defense, particularly distrusted Clarke. Several times Hamre heard complaints from four-star generals, combatant commanders in the field, that Clarke had directly phoned them with orders, rather than going through the secretary of defense, as even the president was supposed to do. Once, Clarke told a general that the president wanted to move a company of soldiers to the Congo during a crisis; Hamre looked into it, and found out the president had asked for no such thing. (Clinton eventually did sign the order, but to Hamre and a number of generals, that didn't excuse Clarke's presumptuousness.)

Hamre's resentment had deeper roots. A few times, when he was the Pentagon's comptroller, he found Clarke raiding the defense budget for "emergency actions," purportedly on behalf of the president. Clarke invoked legal authority for this maneuver—an obscure clause that he'd discovered in the Foreign Assistance Act, Section 506, which allowed the president to take up to $200 million from a department's coffers for urgent, unfunded requirements. Hamre had enough headaches, dealing with post–Cold War budget cuts and pressure from the chiefs, without Clarke swooping down and treating the Pentagon like his piggy bank.

As a result, although they held similar views on several issues, not just cyber security, Hamre hid things from Clarke, sometimes briefing other department deputies in private, rather than in a memo or an NSC meeting, in order to keep Clarke out of the loop.

Around the time of Solar Sunrise and Moonlight Maze, a special prosecutor happened to be investigating charges that President Clinton and the first lady had, years earlier, illegally profited from a land deal in Arkansas. Orders went out from the White House counsel, barring all contact between the White House and the Justice Department, unless it went through him. Clarke ignored the order (he once told an NSA lawyer, "Bureaucrats and lawyers just get in the way") and kept calling the FBI task force for information on its investigation of the hackings. Louis Freeh, the bureau's director, who didn't like Clarke either, told his underlings to ignore the calls.

But Clarke had protectors who valued his advice and gumption. When one agency head urged Sandy Berger, the national security adviser, to fire Clarke, Berger replied, "He's an asshole, but he's my asshole." The president liked that Clarke was watching out for him, too.

Midlevel staffers were simply amazed by the network that Clarke had woven throughout the bureaucracy and by his assertiveness in running it. Once, shortly after coming over from the NSA to be Vice President Gore's intelligence adviser, Rich Wilhelm sat in on a meeting of the NSC counterterrorism subgroup, which Clarke chaired. High-ranking officers and officials, from all the relevant agencies and departments, were at the table, and there was Clarke, this unelected, unconfirmed civilian, barking out orders to an Air Force general to obtain an unmarked airplane and telling the CIA how many agents should board it, all with unquestioned authority.

An aide to Clarke named John McCarthy, a Coast Guard commander with a background in emergency management, attended a Saturday budget meeting, early on in his tenure, where Clarke, upon hearing that an important program fell $3 million short of its needs, told McCarthy to get the money from a certain person at the General Services Administration, adding, "Do it on Monday because I need it on Tuesday." The GSA official told McCarthy he'd give him $800,000, at which point the bargaining commenced. Clarke wound up getting nearly the full sum.

When Clarke replaced Rand Beers, the NSC deputies had been drafting the presidential directive on the protection of critical infrastructure, going back and forth on disagreements and compromise language. Clarke took their work, went back to his office, and wrote the draft himself. It was a detailed document, creating several forums for private-public cooperation on cyber security, most notably Information Sharing and Analysis Centers, in which the government would provide its expertise—including, in some cases, classified knowledge—to companies in the various sectors of critical infrastructure (banking, transportation, energy, and so forth), so they could fix their vulnerabilities.

According to the directive, as Clarke wrote it, this entire effort would be chaired by a new, presidentially appointed official—the

"National Coordinator for Security, Infrastructure Protection, and Counter-terrorism." Clarke made sure, in advance, that he would be this national coordinator.

His detractors, and some of his admirers, saw this as a blatant power grab: he already had the counterterrorism portfolio; now he'd be in charge of critical infrastructure, too. Some of his critics, especially in the FBI, saw it as a substantively bad idea, to boot: cyber threats came mainly from nation-states and criminals; tying the issue to counterterrorism would misconstrue the problem and distract attention from serious solutions. (The idea also threatened to sideline the FBI, which, in the Solar Sunrise and Moonlight Maze investigations, had taken a front-and-center role.)

Clarke waved away the accusations. First, as was often the case, he regarded himself as the best person for the job: he knew more about the issues than anyone else in the White House; ever since the problem had arisen, he and Beers were the only ones to give it more than scant attention. Second, his meetings with Mudge convinced him—he hadn't considered the notion before—that a certain kind of terrorist *could* pull off a devastating cyber attack; it made sense, he coolly explained to anyone who asked, to expand his portfolio in this direction.

As usual, Clarke got his way.

But his directive hit an obstacle with private industry. In Section 5 of PDD-63, laying down "guidelines," Clarke wrote: "The Federal Government shall serve as a model to the private sector on how infrastructure assurance is best achieved and shall, to the extent feasible, distribute the results of its endeavors."

This is what the corporate executives most feared: that the government would be running the show; more to the point, that they would be saddled with the nastiest word in their dictionary—*regulations*. They'd sensed the same threat when they met with the Marsh Commission: here was an Air Force general—and, though retired, he referred to himself as *General* Marsh—laying down the rules on what they must do, as if they were enlisted men. And now here was Dick Clarke, writing under the president's signature, trying to lay down the law.

For several months now, these same companies *had been* working in concert with Washington, under its guidelines, to solve the Y2K crisis. This crisis—also known as the Millennium Bug—emerged when someone realized that some of the government's most vital computer programs had encoded years (dates of birth, dates of retirement, payroll periods, and so forth) by their last two digits: 1995 as "95," 1996 as "96," and so forth. When the calendar flipped to 2000, the computers would read it as "00," and the fear was that they'd interpret it as the year 1900, at which point, all of a sudden, such programs as Social Security and Medicare would screech to a halt: the people who'd been receiving checks would be deemed ineligible because, as far as the computers could tell, they hadn't yet been born. Paychecks for government employees, including the armed forces, could stall; some critical infrastructure, with time-coded programs, might also break down.

To deal with the problem, the White House set up a national information coordination center to develop new guidelines for software and to make sure everyone was on the same page. The major companies, such as AT&T and Microsoft, were brought into the same room with the FBI, the Defense Department, the General Services Administration, the NSA—all the relevant agencies. But the corporate executives made clear that this was a one-time deal; once the Y2K problem was solved, the center would be dismantled.

Clarke wanted to make the arrangement permanent, to turn the Y2K center into the agency that handled cyber threats. Sometime earlier, he'd made no secret of his desire to impose mandatory requirements on cyber security for critical infrastructure, knowing that the private companies wouldn't voluntarily spend the money to

take the necessary actions. But Clinton's economic advisers strenu-
ously opposed the idea, arguing that regulations would distort the
free market and impede innovation. Clinton agreed; Clarke backed
down. Now he was carving a back door, seeking to establish government control through a revamped version of the Y2K center. That
was his agenda in taking over the drafting of the presidential directive—and the companies weren't buying it.

Their resistance put Clarke in a bind. Short of imposing strict requirements, which the president had already struck down, he needed
private industry onboard to make any cyber security policy work: the
vast majority of government data, including a lot of classified data,
flowed through privately controlled networks; and, as the Marsh
Report had shown, the vulnerability of private entities—the critical
infrastructures—had grave implications for national security.

Clarke also knew that, even if the government did take control of
Internet traffic, few agencies possessed the resources or the technical
talent to do much with it—the exceptions being the Defense Department, which had the authority only to defend its own networks, and
the NSA, which had twice been excluded from any role in monitoring civilian computers or telecommunications: first, back in 1984,
in the aftermath of Ronald Reagan's NSDD-145; and, again, early
on in the Clinton presidency, during the Clipper Chip controversy.

Clarke spent much of the next year and a half, in between various
crises over terrorism, writing a 159-page document called the *National Plan for Information Systems Protection: Defending America's Cyberspace*, which President Clinton signed on January 7, 2000.

In an early draft, Clarke had proposed hooking up all civilian government agencies—and, perhaps, eventually critical infrastructure
companies—to a Federal Intrusion Detection Network. FIDNET,
as he called it, would be a parallel Internet, with sensors wired to
some government agency's monitor (which agency was left unclear).
If the sensors detected an intrusion, the monitor would automatically

be alerted. FIDNET would unavoidably have a few access points to the regular Internet, but sensors would sit atop those points and alert officials of intrusions there, as well. Clarke modeled the idea on the intrusion-detection systems installed in Defense Department computers in the wake of Solar Sunrise. But that was a case of the military monitoring itself. To have the government—and, given what agencies did this sort of thing, it would probably be the military—monitoring civilian officials, much less private industry, was widely seen, and loathed, as something different.

When someone leaked Clarke's draft to *The New York Times*, in July 1999, howls of protest filled the air. Prominent members of Congress and civil-liberties groups denounced the plan as "Orwellian." Clarke tried to calm these fears, telling reporters that FIDNET wouldn't infringe on individual networks or privacy rights in the slightest. Fiercer objections still came from the executives and board members of the infrastructure companies, who lambasted the plan as the incarnation of their worst nightmares about government regulation.

The idea was scuttled; the *National Plan* was rewritten.

When the revision was finished and approved six months later, President Clinton scrawled his signature under a dramatic cover note, a standard practice for such documents. But, in a departure from the norm, Clarke—under his own name—penned a separate introduction, headlined, "Message from the National Coordinator."

In it, he tried to erase the image of his presumptuousness. "While the President and Congress can order Federal networks to be secured," he wrote, "they cannot and should not dictate solutions for private sector systems," nor will they "infringe on civil liberties, privacy rights, or proprietary information." He added, just to make things clearer, that the government "will eschew regulation."

Finally, in a gesture so conciliatory that it startled friends and foes alike, Clarke wrote, "This is Version 1.0 of the Plan. We earnestly seek and solicit views about its improvement. As private sector entities make more decisions and plans to reduce their vulnerabilities and improve their protections, future versions of the Plan will reflect that progress."

Then, one month later, the country's largest online companies—including eBay, Yahoo, and Amazon—were hit with a massive denial-of-service attack. Someone hacked into thousands of their computers, few of which were protected in any way, and flooded them with endless requests for data, overloading the servers to the point where they shut down for several hours, in some cases days.

Here was Clarke's chance to jump-start national policy—if not to revive FIDNET (that seemed out of the question for now), then at least to impose some rules on wayward bureaucracies and corporations. He strode into the Oval Office, where Clinton had already heard the news, and said, "This is the future of e-commerce, Mr. President."

Clinton replied, a bit distantly, "Yeah, Gore's always going on about 'e-commerce.'"

Still, Clarke persuaded the president to hold a summit in the White House Cabinet Room, inviting twenty-one senior executives from the major computer and telecom companies—AT&T, Microsoft, Sun Microsystems, Hewlett-Packard, Intel, Cisco, and others—along with a handful of software luminaries from consulting firms and academia. Among this group was the now-famous Peiter Zatko, who identified himself on the official guest list as "Mudge."

Zatko came into the meeting starstruck, nearly as much by the likes of Vint Cerf, one of the Internet's inventors, as by the president of the United States. But after a few minutes of sitting through the discussion, he grew impatient. Clinton was impressive, asking insightful questions, drawing pertinent analogies, grasping the problem at its core. But the corporate execs were faking it, intoning that the attack had been "very sophisticated" without acknowledging that their own passivity had allowed it to happen.

A few weeks earlier, Mudge had gone legit. The L0pht was purchased by an Internet company called @stake, which turned the Watertown warehouse into a research lab for commercial software to block viruses and hackers. Still, he had no personal stake in the piece of theater unfolding before him, so he spoke up.

"Mr. President," he said, "this attack was *not* sophisticated. It was trivial." All the companies should have known that this could happen, but they hadn't invested in preventive measures—which were readily available—because they had no incentive to do so. He didn't elaborate on the point, but everyone knew what he meant by "incentives": if an attack took place, no one would get punished, no stock prices would tank, and it would cost no more to repair the damage than it would have cost to obstruct an attack in the first place.

The room went silent. Finally, Vint Cerf, the Internet pioneer, said, "Mudge is right." Zatko felt flattered and, under the circumstances, relieved.

As the meeting broke up, with everyone exchanging business cards and chatting, Clarke signaled Zatko to stick around. A few minutes later, the two went into the Oval Office and talked a bit more with the president. Clinton admired Zatko's cowboy boots, hoisted his own snakeskins onto his desk, and disclosed that he owned boots made of every mammal on the planet. ("Don't tell the liberals," he whispered.) Zatko followed the president's lead, engaging in more small talk. After a few minutes, a handshake, and a photo souvenir, Zatko bid farewell and walked out of the office with Clarke.

Zatko figured the president had enough on his mind, what with the persistent fallout from the Monica Lewinsky scandal (which had nearly led to his ouster), the fast-track Middle East peace talks (which would go nowhere), and the upcoming election (which Vice President Gore, the carrier of Clinton's legacy, would lose to George W. Bush).

What Zatko didn't know was that, while Clinton could muster

genuine interest in the topic—or any other topic—at a meeting of high-powered executives, he didn't care much about cyber and, really, never had. Clarke was the source, and usually the only White House source, of any energy and pressure on the issue.

Clarke knew that Zatko's Cabinet Room diatribe was on the mark. The industry execs would never fix things voluntarily. In this sense, the meeting was almost comical, with several of them imploring the president to take action, then, a moment later, assuring him that they could handle the problem without government fiat.

The toned-down version of his *National Plan for Information Systems Protection* called for various cooperative ventures between the government and private industry to get under way by the end of 2000 and to be fully in place by May 2003. But the timetable seemed implausible. The banks were game; a number of them had readily agreed to form an industry-wide ISAC—an Information Sharing and Analysis Center—to deal with the challenge. This wasn't so surprising: banks had been the targets of dozens of hackings, costing them millions of dollars and, potentially, the trust of high-rolling customers; some of the larger financial institutions had already hired computer specialists. But most of the other critical infrastructures— transportation, energy, water supply, emergency services—hadn't been hacked: executives of those companies saw the threat as hypothetical; and, as Zatko had observed, they saw no incentive in spending money on security.

Even the software industry included few serious takers: they knew that security was a problem, but they also knew that installing truly secure systems would slow down a server's operations, at a time when customers were paying good money for more speed. Some executives asked security advocates for a cost-benefit analysis: what were the odds of a truly catastrophic event; what would such an event cost them; how much would a security system cost, and what were the chances that the system would actually prevent intrusions?

No one could answer these questions; there were no data to support an honest answer.

The Pentagon's computer network task force was facing similar obstacles. Once, when Art Money, the assistant secretary of defense for command, control, communications, and intelligence, pushed for a 10 percent budget hike for network security, a general asked him whether the program would yield a 10 percent increase in security. Money went around to his technical friends, in the NSA and elsewhere, posing the question. No one could make any such assurance. The fact was, most generals and admirals wanted more tanks, planes, and ships; a billion dollars more for staving off computer attacks—a threat that most regarded as far-fetched, even after Eligible Receiver, Solar Sunrise, and Moonlight Maze (because, after all, they'd done no discernible damage to national security)—meant a billion dollars less for weapons.

But things were changing on the military side: in part because more and more colonels, even a few generals, *were* starting to take the problem seriously; in part because the flip side of cyber security—cyber warfare—was taking off in spades.

CHAPTER 7

DENY, EXPLOIT, CORRUPT, DESTROY

B ACK in the summer of 1994, while Ken Minihan and his demon-dialers at Kelly Air Force Base were planning to shut
down Haiti's telephone network as a prelude to President Clinton's
impending invasion, a lieutenant colonel named Walter "Dusty"
Rhoads was sitting in a command center in Norfolk, Virginia, waiting for the attack to begin.

Rhoads was immersed in Air Force black programs, having started
out as a pilot of, first, an F-117 stealth fighter, then of various experimental aircraft in undisclosed locations. By the time of the Haiti
campaign, he was chief of the Air Combat Command's Information
Warfare Branch at Nellis Air Force Base, Virginia, and, in that role,
had converted Minihan's phone-jamming idea into a detailed plan
and coordinated it with other air operations.

For days, Rhoads and his staff were stuck in that office in Norfolk,
going stir-crazy, pigging out on junk food, while coining code words
for elaborate backup plans, in case one thing or another went wrong.
The room was strewn with empty MoonPie boxes and Fresca cans,

so he made those the code words: "Fresca" for Execute the war plan, "MoonPie" for Stand down.

After the Haitian putschists fled and the invasion was canceled, Rhoads realized that the setup had been a bit convoluted. He was working through Minihan's Air Force Information Warfare Center, which was an intelligence shop, not an operations command; and, strictly speaking, intel and combat ops were separate endeavors, with Title 10 of the U.S. Code covering combat and Title 50 covering intelligence. Rhoads thought it would be a good idea to form an Air Force operations unit dedicated to information warfare.

Minihan pushed for the idea that fall, when he was reassigned to the Pentagon as the assistant chief of staff for intelligence. He sold the idea well. On August 15, 1995, top officials ordered the creation of the 609th Air Information Warfare Squadron, to be located at Shaw Air Force Base, in South Carolina.

The official announcement declared that the squadron would be "the first of its kind designed to counter the increasing threat to Air Force information systems." But few at the time took any such threat seriously; the Marsh Report, Eligible Receiver, Solar Sunrise, and Moonlight Maze wouldn't dot the landscape for another two years. The squadron's other, main mission—though it was never mentioned in public statements—was to develop ways to threaten the information systems of America's adversaries.

Rhoads would be the squadron's commander, while its operations officer would be a major named Andrew Weaver. The previous spring, Weaver had written an Air Staff pamphlet called *Cornerstones of Information Warfare*, defining the term as "any action to deny, exploit, corrupt, or destroy the enemy's information and its functions," with the ultimate intent of "degrading his will or capability to fight." Weaver added, by way of illustration, "Bombing a telephone switching facility is information warfare. So is destroying the switching facility's software."

On October 1, the 609th was up and running, with a staff of just three officers—Rhoads, Weaver, and a staff assistant—occupying a tiny room in the Shaw headquarters basement, just large enough for three desks, one phone line, and two computers.

Within a year, the staff grew to sixty-six officers. Two thirds of them worked on the defensive side of the mission, one third on offense. But in terms of time and energy, the ratio was reversed—one third was devoted to defense, two thirds to offense—and those working the offensive side were kept in separate quarters, behind doors with combination locks.

In February 1997, the squadron held its first full Blue Flag exercise. The plan was for the offensive crew to mount an information warfare attack on Shaw's air wing, while the defensive crew tried to blunt the attack. One of the air wing's officers scoffed at the premise: the wing's communications were all encrypted, he said; nobody can get in there.

But the aggressors broke the passwords, sniffed out the network, found holes, burrowed through, and, once inside, took control. They issued false orders to lighten the air wing's weapons loads, so that the planes would inflict less damage against the enemy. They altered the routes and schedules of tanker aircraft, which were supposed to refuel fighter jets in midflight, as a result of which the fighters ran out of gas before they could carry out their missions.

It was a tabletop game, not a live-action exercise; but if the game had been real, if a wartime adversary had done what the aggressors of the 609th did, the U.S. Air Force's war plan would have been wrecked. Some pilots, looking at their orders, might have realized something was amiss, and made adjustments, but from that point on, neither they nor their commanders would have known whether they could trust *any* orders they received or any information they saw or heard; they would have lost confidence in their command-control.

Toward the end of the game, following a canned script, the defense staved off the attack on the wing's information systems and prevailed in battle. But in fact, everyone knew that the game was a rout in the opposite direction. If the aggressors hadn't been limited by the game's set of rules, they could have shut down the wing's entire operations. Just as Eligible Receiver would demonstrate a few months later, on a wider playing field, the U.S. military—in this case, a vital wing of the Air Force—was horribly vulnerable to an information warfare attack and unable to do anything about it.

Rhoads knew how to shut down the air wing in the Blue Flag exercise because, back when he was chief of the Air Combat Command's Information Warfare Branch, he'd used some of these same techniques in simulations of attacks on *enemy* air wings.

A few months after the Blue Flag demonstration, a real war broke out, and the new commanders of information warfare made their combat debut, better-positioned and higher-ranked than they'd been in the war against Saddam Hussein at the start of the decade.

————

For the previous year, the United States and its NATO allies had been enforcing the Dayton Accords—the December 1995 treaty ending Serbian president Slobodan Milosevic's brutal war in Bosnia-Herzegovina—through an organization called the Stabilization Force, or SFOR, which was also hunting down Serbian war criminals and striving to ensure that the country's elections, scheduled for September 1977, were free and fair.

SFOR had a "white" side, consisting of regular armed forces, and a "black" side, consisting of special-ops units and spies. The black side needed some help; Milosevic wasn't cracking down on war criminals, as he'd promised. So it turned to J-39, Soup Campbell's ultrasecret unit in the Pentagon's Joint Staff that—through links with the NSA, the 609th Information Warfare Squadron, the Air Force Information Warfare Center in San Antonio, and other intelligence agencies—developed the tools and techniques for what they saw as the new face of combat.

J-39 got its first taste of action on July 10, 1997, with Operation Tango, in which five-man teams of British special-ops forces, pretending to be Red Cross officials, captured four of the most-wanted Serbian war criminals. The operation had been preceded by covert surveillance ops—tapping phones, tagging cars with GPS transmitters, and, in a few key areas, installing cameras inside objects that looked like rocks (a contraption designed by Army intelligence technicians at Fort Belvoir, Virginia).

At its peak, more than thirty thousand NATO troops took part in SFOR, a high-profile deployment by any measure, prompting Serbian citizens to mount frequent demonstrations against the Westerners' presence. American officials soon realized that the protests were orchestrated by certain local TV newscasters, who told viewers to go to a specific location, at a specific time, and throw rocks at Western soldiers.

Eric Shinseki, the U.S. Army general in charge of NATO forces in Bosnia, asked the Joint Staff—which, in turn, ordered J-39—to devise some way of turning off TV transmitters when these newscasts came on the air.

Some of the J-39 technicians were from Texas and knew of remote-control devices used at oil wells to turn the pumps off and on. They contracted Sandia Laboratories, a high-tech defense firm, to build a similar device for this operation. Meanwhile, analysts at Kelly Air Force Base calculated that just five television towers were transmitting broadcasts to eighty-five percent of Serbian homes. Some Serbs, who were secretly working for SFOR's black section, installed Sandia's boxes on those five transmitters. Where agents couldn't install them covertly, they told a guard that the box was a new filter for higher-resolution video quality; the guard waved them through.

Once the boxes were set up, engineers at SFOR headquarters

monitored the TV stations. Whenever a newscaster started urging viewers to go demonstrate, they turned off the transmitter carrying that channel's signals.

American officials also drew on their connections to Hollywood, persuading a few TV producers to provide popular programs to the one friendly local station. During the hours when demonstrations were frequently held, the station would run episodes of *Baywatch*, the most popular show in the world; many Serbs, who might otherwise have hit the streets to make trouble, stayed in to watch young women cavorting in bikinis.

General Shinseki visited headquarters for a demonstration of this technology. He asked the engineer who was monitoring the stations to turn off one of the transmission sites. The engineer flicked a switch, and the stations carried by that tower went dead.

Shinseki was amazed. One of the engineers, watching the general's reaction, rolled his eyes and whispered to a colleague, "C'mon, it's an *on-off switch!*"

This wasn't the most sophisticated stunt the team was capable of pulling.

A few months later, it was clear the Dayton Accords were breaking down. General Wesley Clark, the NATO commander, started planning air strikes against Milosevic's key military targets. The J-39 unit laid the groundwork well ahead of time.

The first step of any bombing run would be to disrupt or disable the enemy air-defense system. Two specialists, on loan from a special intelligence unit in Arizona, discovered that Serbia's air-defense system ran through the country's civilian telecommunications system. (Echoes of the aborted 1994 invasion of Haiti, when demon-dialers at Kelly Air Force Base learned the same thing about that country and planned to turn off the radar by flooding the entire phone system with busy signals.)

With the permission of Secretary of Defense William Cohen (who needed to approve any offensive operation involving information warfare), the J-39 unit—which had its own share of former demon-dialers—hacked into the Serbian phone system to scope out everything that General Clark and his planning staff might need to know: how it operated, where it was vulnerable.

The hack was enabled by two bits of good timing. First, CIA director George Tenet had recently created a clandestine unit called the IOC, the Information Operations Center, the main purpose of which was to send in spies to plant a device—a wiretap, a floppy disk, in later years a thumb drive, or whatever else it might take—that would allow SIGINT teams at the NSA or some other agency to intercept communications. In this instance, IOC installed a device at the Serbian phone company's central station.

The other bit of luck was that the Serbs had recently given their phone system a software upgrade. The Swiss company that sold them the software gave U.S. intelligence the security codes.

Once the J-39 tech crews had broken into the Serbian phone system, they could roam through the entire network—including the air-defense lines and telecommunications for the entire Serbian military.

A U.S. Army colonel, monitoring the operation back in the Pentagon, briefed John Hamre, the deputy secretary of defense, on what was going on. Hamre asked how much confidence he had that the plan would frustrate the Serbian commanders.

The colonel replied, "Based on my experience as a battalion commander, if you pick up a phone and can't hear or talk to anyone, it's very frustrating."

"That's good enough for me," Hamre said.

General Clark began the NATO bombing campaign on March 24, 1999. Air Force commanders didn't trust the clever radar-spoofing scheme and instead ordered pilots to fly at very high altitudes, at

least fifteen thousand feet, beyond the range of Serbian anti-air missiles. But on the few occasions when allied planes did dip low, J-39's
operators hacked into the air defense system as planned, and fed it
false information, making the radar screen monitors think the planes
were coming from the west, when in fact they were coming from the
northwest.

The deception had to be subtle; the radar had to be just a bit off,
enough to make Serbian officers blame the miss on a mechanical
flaw but not enough for them to suspect sabotage, in which case
they might switch from automatic guidance to manual control. (The
Serbs managed to shoot down two planes in the course of the war, an
F-16 jet and an F-117 stealth fighter, when an officer made precisely
that switch.) Otherwise, the air-defense crews kept aiming their
weapons at swaths of the sky where no planes were flying.

Another goal of J-39's campaign was to drive a wedge between
Milosevic's paramilitary forces (known as the MUP) and the regular
Yugoslav military (the VJ). The NSA had obtained phone and fax
numbers for officers in both organizations. J-39 officers sent messages to the VJ leaders, expressing admiration for their profession-
alism in defending the Yugoslav people and urging them to remain
apolitical. At one point, General Clark bombed the MUP and VJ
headquarters at roughly the same time. While the planes were in
flight, J-39 sent a message to VJ leaders, warning them to get out of
the building. After both structures were destroyed, the MUP survivors—some of them injured, all of them shaken up—heard that the
VJ officers had fled their headquarters ahead of time, unscathed, and
so they began to suspect that VJ was collaborating with NATO. The
distrust tore the two apart, just as J-39 intended.

As J-39 operators dug deeper into the Serbian military's command-control, they started intercepting communications between
Milosevic and his cronies, many of them civilians. Again with the

assistance of the NSA, the information warriors mapped this social network, learning as much as possible about the cronies themselves, including their financial holdings. As one way to pressure Milosevic and isolate him from his power base, they drew up a plan to freeze his cronies' assets.

The Pentagon's lawyers overruled the proposal—in fact, adamantly rejected any plan designed to affect Serbian civilians. But then, over the weekend of April 17, the Belgrade marathon took place, in which runners of the 26.2-mile race twice crossed a bridge that had been a prominent target in the bombing campaign. The Serbian authorities touted the event—on local and international airwaves—as a defiant protest of NATO's air war, proof of the West's craven weakness in the face of the Serbian people's courage and their loyalty to Milosevic.

President Clinton watched a TV broadcast of the marathon in a foul mood. The previous Monday, a federal judge had found him in contempt of court for giving "intentionally false" testimony about his relations with White House intern Monica Lewinsky. And now this! Wes Clark had promised him that Milosevic would fold after a few days of bombing, yet four *weeks* had passed, and the bastard was thumbing his nose at the Western world.

Clinton sent word to step up the pressure. Suddenly the Pentagon lawyers withdrew their objections to go after Milosevic's cronies. J-39 commenced the next phase of operations the following Monday.*

---

* J-39 also figured out how to hack into Milosevic's own bank accounts; President Clinton was intrigued with the idea. But senior officials, especially in the Treasury Department, strongly advised against going down that road, warning of severe blowback. In subsequent years, intelligence agencies tracked down other hostile leaders' finances, but the option of actually hacking their bank accounts was never actively pursued.

One of Milosevic's major political donors owned a copper mine. J-39 sent him a letter, warning that the mine would be bombed if he didn't stop supporting the Serbian president. The donor didn't respond. Not long before, a CIA contractor had invented a device, made from long strands of carbon fiber, that short-circuited electrical wire on contact. An American combat plane flew over the copper mine, dropped the carbon fiber over the mine's power line, and shut off its electricity. The repair was quick and easy, but so was the message. The donor received another letter, saying that the power outage was a warning: if he didn't change his ways, bombs would fall. He instantly cut off contact with Milosevic.

J-39 also stepped up its campaign to shut down Milosevic's propaganda machine. A European satellite company was carrying the broadcasts of some pro-Milosevic stations. A senior officer in U.S. European Command visited the company's chairman and told him that 80 percent of his board members were from NATO nations. When the chairman told him how much the Serbian stations were paying him, the American officer offered to pay a half million dollars more if he shut them down. He complied.

Meanwhile, U.S. intelligence agencies had discovered that Milosevic's children were vacationing in Greece. Spies took photos of them, lying on the beach. After one bombing run that turned off electrical power in Belgrade, American planes dropped leaflets with the photos beneath a headline blaring that Milosevic had sent his kids to sunbathe in Greece while his own people were sitting in the dark.

Finally, J-39 embarked on a campaign to annoy Milosevic and those around him. They rang his home phone over and over, day and night. When someone picked up, they said nothing. The British equivalent of NSA—the Government Communications Headquarters, or GCHQ—monitored the calls and circulated tape recordings of Madame Milosevic cursing and slamming down the phone. One

GCHQ merrily told his American counterpart, "We like it when they talk dirty to us."

The unit also called Milosevic's generals on their home phones and played a recording of someone who identified himself as General Clark, jovially asking, in fluent Serbo-Croatian, how things were going and imploring them to stop fighting.

On June 4, Milosevic surrendered. It was widely observed that no one had ever before won a war through airpower alone. But this war wasn't won that way, either. It was won through a combination of the pummeling air strikes *and* the isolating impact of information warfare.

Afterward, in a postwar PowerPoint briefing, Admiral James Ellis, Commander of Allied Forces, Southern Europe, hailed the information operation as "at once a great success . . . and perhaps the greatest failure of the war." All the tools were in place, he went on, but "only a few were used." The campaign employed "great people" with "great access to leadership," but they hadn't been integrated with the operational commands, so they had less impact "on planning and execution" than they might have had. The whole enterprise of information warfare, Ellis wrote, had "incredible potential" and "must become" a "point of main effort" in the asymmetric wars to come. However, the concept was "not yet understood by war fighters." One reason for this lapse, he said, was that everything about information warfare was "classified beyond their access," requiring special security clearances that only a few officers possessed. Had the tools and techniques been fully exploited, Ellis concluded, the war might have lasted half as long.

This was the most telling aspect of the information warfare campaign: it was planned and carried out by a secret unit of the Pentagon's Joint Staff, with assistance from the even more secretive NSA, CIA, and GCHQ. As the twentieth century came to a close,

America's military commanders weren't yet willing to let hackers do the business of soldiers and bombardiers. A few senior officers were amenable to experimenting, but the Defense Department lacked the personnel or protocols to integrate this new dimension of war into an actual battle plan. The top generals had signed doctrinal documents on "information warfare" (and, before that, "counter command-control war"), but they didn't appear to take the idea very seriously.

A small group of spies and officers set out to change that.

CHAPTER 8

TAILORED ACCESS

Art Money was flustered. He was the ASD(C3I), the assistant secretary of defense for command, control, communications, and intelligence—and thus the Pentagon's point man on information warfare, its civilian liaison with the NSA. The past few years should have vindicated his enthusiasms. Eligible Receiver, Solar Sunrise, and Moonlight Maze had sired an awareness that the military's computer networks were vulnerable to attack. J-39's operations in the Balkans proved that the vulnerabilities of other countries' networks could be exploited for military gain—that knowing how to exploit them could give the United States an advantage in wartime. And yet, few of America's senior officers evinced the slightest interest in the technology's possibilities.

Money's interest in military technology dated back to a night in 1957, when, as a guard at an Army base in California, he looked up at the sky and saw *Sputnik II*, the Soviet Union's second space satellite, orbiting the earth before the Americans had launched even a first—a beacon of the future, at once fearsome and enthralling. Four years later, he enrolled at San Jose State for an engineering degree.

Lockheed's plant in nearby Sunnyvale was hiring any engineer who could breathe. Money took a job on the night shift, helping to build the system that would launch the new Polaris missile from a tube in a submarine. Soon he was working on top secret spy satellites and, after earning his diploma, the highly classified devices that intercepted radio signals from Soviet missile tests.

From there, he went to work for ESL, the firm that Bill Perry had founded to develop SIGINT equipment for the NSA and CIA; by 1990, Money rose to the rank of company president. Six years later, at the urging of Perry, his longtime mentor, who was now secretary of defense, he came to work at the Pentagon, as assistant secretary of the Air Force for research, development, and acquisition.

That job put him in frequent touch with John Hamre, the Pentagon's comptroller. In February 1998, Solar Sunrise erupted; Hamre, now deputy secretary of defense, realized, to his alarm, that no one around him knew what to do; so he convinced his boss, Secretary of Defense William Cohen, to make Art Money the new ASD(C3I).

Money was a natural for the job. Hamre was set on turning cyber security into a top priority; Money, one of the Pentagon's best-informed and most thoroughly connected officials on cyber matters, became his chief adviser on the subject. It was Money who suggested installing intrusion-detection systems on Defense Department computers. It was Money who brought Dusty Rhoads into J-39 after hearing about his work in the Blue Flag war games at the 609th Information Warfare Squadron. It was Money who brought together J-39, the NSA, and the CIA during the campaign in the Balkans.

The concept of information warfare—or cyber warfare, as it was now called—should have taken off at this point, but it hadn't because most of the top generals were still uninterested or, in some cases, resistant.

In the summer of 1998, in the wake of Solar Sunrise, Money was instrumental in setting up JTF-CND—Joint Task Force-Computer

Network Defense—as the office to coordinate protective measures for all Defense Department computer systems, including the manning of a 24/7 alert center and the drafting of protocols spelling out what to do in the event of an attack. In short, Money was piecing together the answer to the question Hamre posed at the start of Solar Sunrise: "Who's in charge?"

The initial plan was to give Joint Task Force-Computer Network Defense an *offensive* role as well, a mandate to develop options for attacking an adversary's networks. Dusty Rhoads set up a small, hush-hush outpost to do just that. But he, Money, and Soup Campbell, the one-star general in charge of the task force, knew that the services wouldn't grant such powers to a small bureau with no command authority.

However, Campbell made a case that, to the extent the military services had plans or programs for cyber offensive operations (and he knew they did), the task force ought, at the very least, to be briefed on them. His argument was unassailable: the task force analysts needed to develop defenses against cyber attacks; knowing what kinds of attacks the U.S. military had devised would help them expand the range of defenses—since, whatever America was plotting against its adversaries, its adversaries would likely soon be plotting against America.

Cohen bought the argument and wrote a memo to the service chiefs, ordering them to share their computer network attack plans with the joint task force. Yet at a meeting chaired by John Hamre, the vice chiefs of the Army, Navy, and Air Force—speaking on behalf of their bosses—blew the order off. They didn't explicitly disobey the order; that would have been insubordination, a firing offense. Instead, they redefined their attack plans as something else, so they could say they had no such plans to brief. But their evasion was obvious: they just didn't want to share these secrets with others, not even if the secretary of defense told them to do so.

Clearly, the task force needed a broader charter and a home with more power. So, on April 1, 2000, JTF-CND became JTF-CNO, the O standing for "Operations," and those operations included not just Computer Network Defense but also, explicitly, Computer Network *Attack*. The new task force was placed under the purview of U.S. Space Command, in Colorado Springs. It was an odd place to be, but SpaceCom was the only unit that wanted the mission. In any case, it was a *command*, invested with war-planning and war-fighting powers.

Still, Money, Campbell, Hamre, and the new task force commander, Major General James D. Bryan, saw this, too, as a temporary arrangement. Colorado Springs was a long way from the Pentagon or any other power center; and the computer geeks from the task force were complaining that their counterparts at Space Command, who had to be meshed into the mission, didn't know anything about cyber offense.

Money felt that the cyber missions—especially those dealing with cyber *offense*—should ultimately be brought to the Fort Meade headquarters of the NSA. And so did the new NSA director, Lieutenant General Michael Hayden.

————

Mike Hayden came to the NSA in March 1999, succeeding Ken Minihan. It wasn't the first time Hayden followed in his footsteps. For close to two years, beginning in January 1996, Hayden commanded Kelly Air Force Base in San Antonio. Kelly was where Minihan had run the Air Force Information Warfare Center, which pioneered much of what came to be called cyber warfare—offense *and* defense—and, by the time Hayden arrived, it had grown in sophistication and stature.

Hayden knew little about the subject before his tenure at Kelly, but he quickly realized its possibilities. A systematic thinker who liked to place ideas in categories, he came up with a mission concept that he called GEDA—an acronym for Gain (collect information), Exploit (use the information to penetrate the enemy's networks), Defend (prevent the enemy from penetrating our networks), Attack (don't just penetrate the enemy network—disable, disorient, or destroy it).

At first glance, the concept seemed obvious. But Hayden's deeper point was that all these missions were intertwined—they all involved the same technology, the same networks, the same actions: intelligence and operations in cyberspace—cyber security, cyber espionage, and cyber war—were, in a fundamental sense, synonymous.

Hayden was stationed overseas, as the intelligence chief for U.S. forces in South Korea, when Solar Sunrise and Moonlight Maze stirred panic in senior officialdom and made at least some generals realize that the trendy talk about "information warfare" might be worthy of attention. Suddenly, if just to stake a claim in upcoming budget battles, each of the services hung out a cyber shingle: the Army's Land Information Warfare Activity, the Navy's Naval Information Warfare Activity, and even a Marine Corps Computer Network Defense unit, joined the long-standing Air Force Information Warfare Center in the enterprise.

Many of these entities had sprung up during Ken Minihan's term as NSA director, and the trend worried him for three reasons. First, there were financial concerns: the defense budget was getting slashed in the wake of the Cold War; the NSA's share was taking still deeper cuts; and he didn't need other, more narrowly focused entities—novices in a realm that the NSA had invented and mastered—to drain his resources further. Second, some of these aspiring cyber warriors had poor operational security; they were vulnerable to hacking by adversaries, and if an adversary broke into their networks, he might gain access to files that the NSA had shared.

Finally, there was an existential concern. When Minihan became

NSA director, Bill Perry told him, "Ken, you need to preserve the mystique of Fort Meade." The *mystique*—that was the key to the place, Minihan realized early on: it was what swayed presidents, cabinet secretaries, committee chairmen, and teams of government lawyers to let the NSA operate in near-total secrecy, and with greater autonomy than the other intelligence agencies. Fort Meade was where brilliant, faceless code-makers and code-breakers did things that few outsiders could pretend to understand, much less duplicate; and, for nearly the entire post–World War II era, they'd played a huge, if largely unreported, role in keeping the peace.

Now, the mystique was unraveling. With the Cold War's demise, Minihan gutted the agency's legendary A Group, the Soviet specialists, in order to devote more resources to emerging threats, including rogue regimes and terrorists. The agency could still boast of its core *technical* base: the cryptologists, the in-house labs, and their unique partnership with obscure outside contractors—that was where the mystique still glowed. Minihan needed to build up that base, expand its scope, shift its agenda, and preserve its mastery—not let it be diluted by lesser wannabes splashing in the same stream.

Amid the profusion of entities claiming a piece of Fort Meade's once-exclusive turf, and the parallel profusion of terms for what was essentially the same activity ("information warfare," "information operations," "cyber warfare," and so forth), Minihan tried to draw the line. "I don't care what you call it," he often said to his political masters. "I just want you to call *me*."

To keep NSA at the center of this universe, Minihan created a new office, at Fort Meade, called the IOTC—the Information Operations Technology Center. The idea was to consolidate all of the military's sundry cyber shops: not to destroy them—he didn't want to set off bureaucratic wars—but to corral them into his domain.

He had neither the legal authority nor the political clout to do this by fiat, so he asked Art Money, whom he'd known for years and who'd just become ASD(C3I), to scour the individual services' cyber budgets for duplicative programs; no surprise, Money found many. He took his findings to John Hamre, highlighted the redundancies, and made the pitch. No agency, Money said, could perform these tasks better than the NSA—which, he added, happened to have an office called the IOTC, which would be ideal for streamlining and coordinating these far-flung efforts. Hamre, who had recently come to appreciate the NSA's value, approved the idea and put the new center under Money's supervision.

When Hayden took over NSA, Money pressed him to take the center in a different direction. Minihan's aim, in setting up the IOTC, was to emphasize the *T*—Technology: that was the NSA's chief selling point, its rationale for remaining at the top of the pyramid. Money wanted to stress the *O*—Operations: he wanted to use the IOTC as a back door for the NSA to get into cyber offensive operations.

The idea aroused controversy on a number of fronts. First, within the NSA, many of the old-timers didn't like it. The point of SIGINT, the prime NSA mission, was to gather intelligence by penetrating enemy communications; if the NSA *attacked* the source of those communications, then the intelligence would be blown; the enemy would know that we knew how to penetrate his network, and he'd change his codes, revamp his security.

Second, Money's idea wasn't quite legal. Broadly, the military operated under Title 10 of the federal statutes, while the intelligence agencies, including the NSA, fell under Title 50. Title 10 authorized the use of force; Title 50 did not. The military could use intelligence gathered by Title 50 agencies as the *basis* for an attack; the NSA could not launch attacks on its own.

Money and Hayden thought a case could be made that the IOTC maneuvered around these strictures because, formally, it reported to the secretary of defense. But the legal thicket was too dense for such a simple workaround. Each of the military services would have a stake in any action that the IOTC might take, as would the CIA and possibly other agencies. It was a ramshackle structure from the start. It made sense from a purely technical point of view: as Minihan and Hayden both realized, from their tenures at the Air Force Information Warfare Center in San Antonio, computer network offense and defense were operationally the same—but the *legal* authorities were separate.

Hayden considered the IOTC a good enough arrangement for the moment. At least, as Minihan had intended, it protected the NSA's dominance in the cyber arena. Expanding its realm over the long haul would have to wait. Meanwhile, Hayden faced a slew of other, daunting problems from almost the moment he took office.

Soon after his arrival at Fort Meade, he got wind of a top secret report, written a few months earlier for the Senate Select Committee on Intelligence, titled "Are We Going Deaf?" It concluded that the NSA, once on the cutting edge of SIGINT technology, had failed to keep pace with the changes in global telecommunications; that, while the world was shifting to digital cell phones, encrypted email, and fiber optics, Fort Meade remained all too wedded to tapping land lines, analog circuits, and intercepting radio frequency transmissions.

The report was written by the Technical Advisory Group, a small panel of experts that the Senate committee had put together in 1997 to analyze the implications of the looming digital age. Most of the group's members were retired NSA officials, who had urged their contacts on the committee to create the advisory group precisely because they were disturbed by Fort Meade's recalcitrant ways and thought that outside prodding—especially from the senators who held its purse strings—might push things forward.

One of the group's members, and the chief (though unnamed) author of this report, was former NSA director Bill Studeman. A full decade had passed since Studeman, upon arriving at Fort Meade, commissioned two major studies: one, projecting how quickly the world would shift from analog to digital; the other, concluding that the skill sets of NSA personnel were out of whack with the requirements of the impending new world.

In the years since, Studeman had served as deputy director of the CIA, joined various intelligence advisory boards, and headed up projects on surveillance and information warfare as vice president of Northrop Grumman Corporation. In short, he was still plugged in, and he was appalled by the extent to which the NSA was broaching obsolescence.

The Senate committee took his report very seriously, citing it in its annual report and threatening to slash the NSA budget if the agency didn't bring its practices up to date.

Studeman's report was circulated while Minihan was still NSA director, and it irritated him. He'd instigated a lot of reforms already; the agency had come a long way since the Senate committee discovered that it was spending only $2 million a year on projects to penetrate the Internet. But he didn't speak out against the report: if the senators believed it, maybe they'd boost the NSA's budget. *That* was his biggest problem: he knew what needed to be done; he just needed more money to do it.

But Hayden, when he took over Fort Meade, took Studeman's report as gospel and named a five-man group of outsiders—senior executives at aerospace contractors who'd managed several intelligence-related projects—to conduct a review of the NSA's organization, culture, management, and priorities. With Hayden's encouragement, they pored over the books and interviewed more than a hundred officials, some inside the NSA, some at other agencies that had dealings—in some cases, contentious dealings—with Fort Meade.

On October 12, after two months of probing, the executives briefed Hayden on their findings, which they, soon after, summarized in a twenty-seven-page report. The NSA, they wrote, suffered from a "poorly communicated mission," a "lack of vision," a "broken personnel system," "poor" relations with other agencies that depended on its intelligence, and an "inward-looking culture," stemming in part from its intense secrecy. As a result of all these shortcomings, NSA managers tended to protect its "legacy infrastructure" rather than develop "new methods to deal with the global network." If it persisted in its outmoded ways, the agency "will fail," and its "stakeholders"—the president, secretary of defense, and other senior officials—"will go elsewhere" for their intelligence.

Few of the group's observations or critiques were new. NSA directors, going back twenty years, had spoken of the looming gap between the agency's tools and the digital world to come. Studeman and his mentor, Bobby Ray Inman, warned of the need to adapt, though too far ahead of time for their words to gain traction. Mike McConnell prodded the machine into motion, but then got caught up in the ill-fated Clipper Chip. Ken Minihan saw the future more clearly than most, but he wasn't a natural manager. He was a good old boy from Texas who dispensed with the formalities of most general officers and played up his down-home style (some called it "his Andy Griffith bit"). Everyone liked him, but few understood what he was talking about. He would drop Air Force aphorisms, like "We're gonna make a hard right turn with our blinkers off," which flew over everyone's head. He'd issue stern pronouncements, like "One team, one mission," but this, too, inspired only uncertainty: he seemed to be saying that someone should work more closely with someone else, but *who* and *with whom*—the SIGINT and Information Assurance Directorates? the NSA and the CIA? the intelligence community and the military? No one quite knew.

Hayden, by contrast, was a modern military general, less brash, certainly less folksy, than Minihan: more of a tight-cornered, chart-sketching manager. As a follow-up to the five-man group's briefing, he circulated throughout the NSA his own eighteen-page, bluntly worded memo titled "The Director's Work Plan for Change," summarizing much of the executives' report and outlining his solutions.

His language was as stark as his message. The NSA, he wrote, "is a misaligned organization," its storied legacy "in great peril." It needed bold new leadership, an integrated workforce in which signals intelligence and information security would act in concert, not at odds with each other (this is what Minihan had meant by "one team, one mission"), and—above all—a refocused SIGINT Directorate that would "address the challenge of technological change."

He concluded, "We've got it backwards. We start with our internal tradecraft, believing that customers will ultimately benefit"—when, in fact, the agency needed to focus first on the needs of the customers (the White House, the Defense Department, and the rest of the intelligence community), then align its tradecraft to those tasks.

Minihan had gone some distance down that road. He decimated the A Group, the Soviet specialists who'd helped win the Cold War, but he didn't erect a structure, or clearly define a new mission worthy of the title A Group in its place. It wasn't entirely his fault: as he frequently complained, he lacked the money, the time, and any direction from his political masters. Ideally, as Hayden would note, the NSA goes out and gets what the nation's leaders want it to get; but no one high up gave Minihan any marching order. Then again, the lack of communication went both ways: no one high up knew what the NSA could offer, apart from the usual goods, which were fine, as far as they went, but they fell short in a world where its tools and techniques were "going deaf."

One of the agency's main problems, according to the aerospace executives' report, was a "broken personnel system." Employees tended to serve for life, and they were promoted through the ranks at the same pace, almost automatically, with little regard for individual talent. This tenured system had obstructed previous stabs at reform: the upper rungs were occupied by people who'd come up in the 1970s and 1980s, when money flowed freely, the enemy was clear-cut, and communications—mainly telephone calls and radio-frequency transmissions—could be tapped by a simple circuit or scooped out of the air.

Hayden changed the personnel system, first of all. On November 15, he inaugurated "One Hundred Days of Change." Before, senior employees wore special badges and rode in special elevators; now, everyone would wear the same badges, and all elevators would be open to all. Hayden also scoured personnel evaluations, consulted a few trusted advisers, and—after the first two weeks—fired sixty people who had been soaking up space for decades and promoted sixty more competent officials, most of them far junior in age and seniority, to fill the vacancies.

Much grumbling ensued, but then, on January 24, 2000, ten weeks into Hayden's campaign, an alarm bell went off: the NSA's main computer system crashed—and stayed crashed for seventy-two hours. The computer was still storing intelligence that the field stations were gathering from all over the world, but no one at Fort Meade could gain access to it. Raw intelligence—unsifted, unprocessed, unanalyzed—was all but useless; for three days, the NSA was, in effect, shut down.

At first, some suspected sabotage or a delayed effect of Y2K. But the in-house tech crews quickly concluded that the computer had simply been overloaded; and the damage was so severe that they'd have to reconstruct the data and programs after it came back online.

The grumbling about Hayden ceased. If anyone had doubted that big changes were necessary, there was no doubt now.

Another criticism in the executives' report was that the SIGINT Directorate stovepiped its data by geography—one group looked at signals from the former Soviet Union, another from the Middle

East, another from Asia—whereas, out in the real world, all communications passed through the same network. The World Wide Web was precisely *that*—worldwide.

In their report, the executives suggested a new organizational chart for the SIGINT Directorate, broken down not along regional lines (which no longer made sense) but rather into "Global Response," "Global Network," and "Tailored Access."

"Global Response" would confront day-to-day crises without diverting resources from the agency's steady tasks. This had been a big source of Minihan's frustrations: the president or secretary of defense kept requesting so much intelligence on one crisis after another—Saddam Hussein's arms buildup, North Korea's nuclear program, prospects for Middle East peace talks—that he couldn't focus on structural reforms.

"Global Network" was the new challenge. In the old days, NSA linguists would sit and listen to live feeds, or stored tapes, of phone conversations and radio transmissions that its taps and antenna dishes were scooping up worldwide. In the new age of cell phones, faxes, and the Internet, there often wasn't anything to *listen* to; and to the extent there was, the signal didn't travel from one point to another, on one line or channel. Instead, digital communications zipped through the network in data *packets*, which were closely interspersed with packets of other communications (a feature that would spark great controversy years later, when citizens learned that the NSA was intercepting *their* conversations as well as those of bad guys). These networks and packets were far too vast for human beings to monitor in real time; the intelligence would have to be crunched, sifted, and processed by very high-speed computers, scanning the data for key words or suspicious traffic patterns.

To Hayden, the three-day computer crash in January suggested that the NSA's own hardware might not be up to the task. The aerospace executives had recommended, with no small self-interest, that the agency should examine what outside contractors might offer. Hayden took them up on their suggestion. New computers and software would be needed to scan and make sense of this new global network; maybe commercial contractors would do a better job of creating them.

He called the new program Trailblazer, and in August, he held Trailblazer Industry Day, inviting 130 corporate representatives to come to Fort Meade and hear his pitch. In October, he opened competition on a contract to build a "technical demonstration platform" of the new system. The following March, the NSA awarded $280 million—the opening allotment of what would amount to more than $1 billion over the next decade—to Science Applications International Corporation, with pieces of the program shared by Northrop Grumman, Boeing, Computer Sciences Corp., and Booz Allen Hamilton, all of which had longtime relations with the intelligence community.

SAIC was particularly intertwined with NSA. Bobby Ray Inman sat on its board of directors. Bill Black, one of the agency's top cryptologists, had retired in 1997 to become the corporation's assistant vice president; then, three years later, in a case of revolving doors that shocked the most jaded insiders, Hayden brought him back in to be the NSA deputy director—and to manage Trailblazer, which he'd been running from the other side of the transom at SAIC.

But the NSA needed a bigger breakthrough still: it needed tools and techniques to intercept signals, not only as they flowed through the digital network but also at their source. The biggest information warfare campaign to date, in the Balkans, had involved hacking into Belgrade's *telephone* system. Earlier that decade, in the Gulf War, when Saddam Hussein's generals sent orders through fiber-optic cable, the Pentagon's Joint Intelligence Committee—which relied heavily on NSA personnel and technology—figured out how to blow up the cable links, forcing Saddam to switch to microwave.

The NSA knew how to intercept microwaves, but it didn't yet know how to intercept the data rushing through fiber optics. That's what the agency now needed to do.

In their report to Hayden, the aerospace executives recommended that the SIGINT and Information Assurance Directorates "work very closely," since their two missions were "rapidly becoming two sides of the same coin."

For years, Information Assurance, located in an annex near Baltimore-Washington International Airport, a half hour's drive from Fort Meade, had been testing and fixing software used by the U.S. military—probing for vulnerabilities that the enemy could exploit. Now one of the main roles of the SIGINT crews, in the heart of the agency's headquarters, was to find and exploit vulnerabilities in the *adversaries'* software. Since people (and military establishments) around the world were using the same Western software, the Information Assurance specialists possessed knowledge that would be valuable to the SIGINT crews. At the same time, the SIGINT crews had knowledge about adversaries' networks—what they were doing, what kinds of attacks they were planning and testing—that would be valuable to the Information Assurance specialists. Sharing this knowledge, on the offense and the defense, required mixing the agency's two distinct cultures.

Inman and McConnell had taken steps toward this integration. Minihan had started to tear down the wall, moving a few people from the annex to headquarters and vice versa. Hayden now widened Minihan's wedge, moving more people back and forth, to gain insights about the security of their own operations.

Another issue that needed to be untangled was the division of labor within the intelligence community, especially between the NSA and the CIA. In the old days, this division was clear: if information moved, the NSA would intercept it; if it stood still, the CIA would send a spy to nab it. NSA intercepted electrons whooshing through the air or over phone lines; CIA stole documents sitting on a desk or in a vault. The line had been sharply drawn for decades. But in the digital age, the line grew fuzzy. Where did computers stand in relation to this line? They stored data on floppy disks and hard drives, which were stationary; but they also sent bits and bytes through cyberspace. Either way, the information was the same, so who should get it: Langley or Fort Meade?

The logical answer was both. But pulling off that feat would require a fusion with little legal or bureaucratic precedent. The two spy agencies had collaborated on the occasional project over the years, but this would involve an institutional melding of missions and functions. To do its part, each agency would have to create a new entity—or to beef up, and reorient, an existing one.

As it happened, a framework for this fusion already existed. The CIA had created the Information Operations Center during the Belgrade operation, to plant devices on Serbian communications systems, which the NSA could then intercept; this center would be Langley's contribution to the new joint effort. Fort Meade's would be the third box on the new SIGINT organizational chart—"tailored access."

Minihan had coined the phrase. During his tenure as director, he pooled a couple dozen of the most creative SIGINT operators into their own corner on the main floor and gave them that mission. What CIA black-bag operatives had long been doing in the physical world, the tailored access crew would now do in cyberspace, sometimes in tandem with the black-baggers, if the latter were needed—as they had been in Belgrade—to install some device on a crucial piece of hardware.

The setup transformed the concept of signals intelligence, the NSA's stock in trade. SIGINT had long been defined as passively collecting stray electrons in the ether; now, it would also involve actively breaking and entering into digital machines and networks.

Minihan had wanted to expand the tailored access shop into an A Group of the digital era, but he ran out of time. When Hayden launched his reorganization, he took the baton and turned it into a distinct, elite organization—the Office of Tailored Access Operations, or TAO.

It began, even under his expansion, as a small outfit: a few dozen computer programmers who had to pass an absurdly difficult exam to get in. The organization soon grew into an elite corps as secretive and walled off from the rest of the NSA as the NSA was from the rest of the defense establishment. Located in a separate wing of Fort Meade, it was the subject of whispered rumors, but little solid knowledge, even among those with otherwise high security clearances. Anyone seeking entrance into its lair had to get by an armed guard, a cipher-locked door, and a retinal scanner.

In the coming years, TAO's ranks would swell to six hundred "intercept operators" at Fort Meade, plus another four hundred or more at NSA outlets—Remote Operations Centers, they were called—in Wahiawa, Hawaii; Fort Gordon, Georgia; Buckley Air Force Base, near Denver; and the Texas Cryptology Center, in San Antonio.

TAO's mission, and informal motto, was "getting the ungettable," specifically getting the ungettable stuff that the agency's political masters wanted. If the president wanted to know what a terrorist leader was thinking and doing, TAO would track his computer, hack into its hard drive, retrieve its files, and intercept its email—sometimes purely through cyberspace (especially in the early days, it was easy to break a target's password, if he'd inscribed a password at all), sometimes with the help of CIA spies or special-ops shadow soldiers, who'd lay their hands on the computer and insert a thumb drive loaded with malware or attach a device that a TAO specialist would home in on.

These devices—their workings and their existence—were so secret that most of them were designed and built inside the NSA: the software by its Data Network Technologies Branch, the techniques by its Telecommunications Network Technologies Branch, and the customized computer terminals and monitors by its Mission Infrastructure Technologies Branch.

Early on, TAO hacked into computers in fairly simple ways: phishing for passwords (one such program tried out every word in the dictionary, along with variations and numbers, in a fraction of a second) or sending emails with alluring attachments, which would download malware when opened. Once, some analysts from the Pentagon's Joint Task Force-Computer Network Operations were invited to Fort Meade for a look at TAO's bag of tricks. The analysts laughed: this wasn't much different from the software they'd seen at the latest DEF CON Hacking Conference; some of it seemed to be repackaged versions of the same software.

Gradually, though, the TAO teams sharpened their skills and their arsenal. Obscure points of entry were discovered in servers, routers, workstations, handsets, phone switches, even firewalls (which, ironically, were supposed to keep hackers out), as well as in the software that programmed, and the networks that connected, this equipment. And as their game evolved, their devices and programs came to resemble something out of the most exotic James Bond movie. One device, called LoudAuto, activated a laptop's microphone and monitored the conversations of anyone in its vicinity. HowlerMonkey extracted and transmitted files via radio signals, even if the computer wasn't hooked up to the Internet. MonkeyCalendar tracked a cell phone's location and conveyed the information through a text message. NightStand was a portable wireless system that loaded a computer with malware from several miles away. RageMaster tapped into a computer's video signal, so a TAO technician could see what was on its screen and thus watch what the person being targeted was watching.

But as TAO matured, so did its targets, who figured out ways to detect and block intruders—just as the Pentagon and the Air Force had figured out ways, in the previous decade, to detect and block intrusions from adversaries, cyber criminals, and mischief-makers. As hackers and spies discovered vulnerabilities in computer software and hardware, the manufacturers worked hard to patch the holes— which prodded hackers and spies to search for new vulnerabilities, and on the race spiraled.

As this race between hacking and patching intensified, practitioners of both arts, worldwide, came to place an enormous value on "zero-day vulnerabilities"—holes that no one had yet discovered, much less patched. In the ensuing decade, private companies would spring up that, in some cases, made small fortunes by finding zero-day vulnerabilities and selling their discoveries to governments, spies, and criminals of disparate motives and nationalities. This hunt for zero-days preoccupied some of the craftiest mathematical minds in the NSA and other cyber outfits, in the United States and abroad.

Once, in the late 1990s, Richard Bejtlich, a computer network defense analyst at Kelly Air Force Base discovered a zero-day vulnerability—a rare find—in a router made by Cisco. He phoned a Cisco technical rep and informed him of the problem, which the rep then quickly fixed.

A couple days later, proud of his prowess and good deed, Bejtlich told the story to an analyst on the offensive side of Kelly. The analyst wasn't pleased. Staring daggers at Bejtlich, he muttered, "Why didn't you tell *us*?"

The implication was clear: if Bejtlich had told the offensive analysts about the flaw, they could have exploited it to hack foreign networks that used the Cisco server. Now it was too late; thanks to Bejtlich's phone call, the hole was patched, the portal was closed.

As the NSA put more emphasis on finding and exploiting vulnerabilities, a new category of cyber operations came into prominence. Before, there was CND (Computer Network *Defense*) and CNA

(Computer Network *Attack*); now there was also CNE (Computer Network *Exploitation*).

CNE was an ambiguous enterprise, legally and operationally, and Hayden—who was sensitive to legal niceties and the precise wiggle room they allowed—knew it. The term's technical meaning was straightforward: the use of computers to *exploit* the vulnerabilities of an adversary's networks—to get inside those networks, in order to gain more intelligence about them. But there were two ways of looking at CNE. It could be the front line of Computer Network Defense, on the logic that the best way to defend a network was to learn an adversary's plans for attack—which required getting inside his network. Or, CNE could be the gateway for Computer Network Attack—getting inside the enemy's network in order to map its passageways and mark its weak points, to "prepare the battlefield" (as commanders of older eras would put it) for an American offensive, in the event of war.*

The concept of CNE fed perfectly into Hayden's desire to fuse cyber offense and cyber defense, to make them indistinguishable. And while Hayden may have articulated the concept in a manner that suited his agenda, he didn't invent it; rather, it reflected an intrinsic aspect of modern computer networks themselves.

In one sense, CNE wasn't so different from intelligence gathering of earlier eras. During the Cold War, American spy planes penetrated the Russian border in order to force Soviet officers to turn on their radar and thus reveal information about their air-defense systems. Submarine crews would tap into underwater cables near Russian ports to intercept communications, and discover patterns, of Soviet naval operations. This, too, had a dual purpose: to bolster defenses

---

*Out of CNE sprang a still more baroque subdivision of signals intelligence: C-CNE, for Counter-Computer Network Exploitation—penetrating an adversary's networks in order to watch him penetrating *our* networks.

against possible Soviet aggression; and to prepare the battlefield (or airspace and oceans) for an American offensive.

But in another sense, CNE was a completely different enterprise: it exposed all society to the risks and perils of military ventures in a way that could not have been imagined a few decades earlier. When officials in the Air Force or the NSA neglected to let Microsoft (or Cisco, Google, Intel, or any number of other firms) know about vulnerabilities in its software, when they left a hole unplugged so they could exploit the vulnerability in a Russian, Chinese, Iranian, or some other adversary's computer system, they also left American citizens open to the same exploitations—whether by wayward intelligence agencies or by cyber criminals, foreign spies, or terrorists who happened to learn about the unplugged hole, too.

This was a new tension in American life: not only between individual liberty and national security (that one had always been around, to varying degrees) but also between different layers and concepts of security. In the process of keeping military networks more secure from attack, the cyber warriors were making civilian and commercial networks *less* secure from the same kinds of attack.

These tensions, and the issues they raised, went beyond the mandate of national security bureaucracies; only political leaders could untangle them. As the twenty-first century approached, the Clinton administration—mainly at the feverish prodding of Dick Clarke—had started to grasp the subject's complexities. There was the Marsh Report, followed by PDD-63, the *National Plan for Information Systems Protection*, and the creation of Information Sharing and Analysis Centers, forums in which the government and private companies could jointly devise ways to secure their assets from cyber attacks.

Then came the election of November 2000, and, as often happens when the White House changes party, all this momentum ground to a halt. When George W. Bush and his aides came to power on January 20, 2001, the contempt they harbored for their predecessors seethed with more venom than usual, owing to the sex scandal and impeachment that tarnished Clinton's second term, compounded by the bitter aftermath of the election against his vice president, Al Gore, which ended in Bush's victory only after the Supreme Court halted a recount in Florida.

Bush threw out lots of Clinton's initiatives, among them those having to do with cyber security. Clarke, the architect of those policies, stayed on in the White House and retained his title of National Coordinator for Security, Infrastructure Protection, and Counterterrorism. But, it was clear, Bush didn't care about any of those issues, nor did Vice President Dick Cheney or the national security adviser, Condoleezza Rice. Under Clinton, Clarke had the standing, even if not the formal rank, of a cabinet secretary, taking part in the NSC Principals meetings—attended by the secretaries of defense, state, treasury, and other departments—when they discussed the issues in his portfolio. Rice took away this privilege. Clarke interpreted the move as not only a personal slight but also a diminution of his issues.

During the first few months of Bush's term, Clarke and CIA director George Tenet, another Clinton holdover, warned the president repeatedly about the looming danger of an attack on America by Osama bin Laden. But the warnings were brushed aside. Bush and his closest advisers were more worried about missile threats from Russia, Iran, and North Korea; their top priority was to abrogate the thirty-year-old Anti-Ballistic Missile Treaty, the landmark Soviet-American arms-control accord, so they could build a missile-defense system. (On the day of the 9/11 attacks, Rice was scheduled to deliver a speech on the major threats facing the land; the draft didn't so much as mention bin Laden or al Qaeda.)

In June 2001, Clarke submitted his resignation. He was the chief White House adviser on counterterrorism, yet nobody was paying attention to terrorism—or to him. Rice, taken aback, urged him not to leave. Clarke relented, agreeing to stay but only if they limited his responsibilities to cyber security, gave him his own staff (which eventually numbered eighteen), and let him set up and run an interagency Cyber Council. Rice agreed, in part because she didn't care much about cyber; she saw the concession as a way to keep Clarke onboard while keeping him out of issues that did interest her. However, she needed time to find a replacement for the counterterrorism slot, so Clarke agreed to stay in that position as well until October 1.

He still had a few weeks to go as counterterrorism chief when the hijacked planes smashed into the World Trade Center and the Pentagon. Bush was in Florida, Cheney was dashed to an underground bunker, and, by default, Clarke sat in the Situation Room as the crisis manager, running the interagency conference calls and coordinating, in some cases directing, the government's response.

The experience boosted his standing somewhat, not enough to let him rejoin the Principals meetings, but enough for Rice to start paying a bit of attention to cyber security. However, she balked when Clarke suggested renewing the *National Plan for Information Systems Protection*, which he'd written for Clinton in his last year as president. She vaguely remembered that the plan set mandatory standards for private industry, and that would be anathema to President Bush.

In fact, much as Clarke wished that it had, the plan—the revised version, after he had to drop his proposal for a federal intrusion-detection network—called only for public-private cooperation, with corporations in the lead. But Clarke played along, agreeing with Rice that the Clinton plan was deeply flawed and that he wanted to do a drastic rewrite. Rice let him draft an executive order, which Bush signed on September 30, calling for a new plan. For the next several months, Clarke and some of his staff went on the road, doing White House "cyber town halls" in ten cities—including Boston, New York, Philadelphia, Atlanta, San Francisco, Los Angeles, Portland,

and Austin—inviting local experts, corporate executives, IT managers, and law-enforcement officers to attend.

Clarke would start the sessions on a modest note. Some of you, he would say, criticized the Clinton plan because you had no involvement in it. Now, he went on, the Bush administration was writing a new plan, and the president wants you, the people affected by its contents, to write the annexes that deal with your sector of critical infrastructure. Some of the experts and executives in some of the cities actually submitted ideas; those in telecommunications were particularly enthused.

In fact, though, Clarke wasn't interested in their ideas. He did, however, need to melt their opposition; the whole point, the only point, of the town hall theatrics was to get their buy-in—to co-opt them into believing that they had something to do with the report. As it turned out, the final draft—a sixty-page document called *The National Strategy to Secure Cyberspace*, signed by President Bush on February 14, 2003—contained more passages kowtowing to industry, and it assigned some responsibility for securing nonmilitary cyberspace to the new Department of Homeland Security. But otherwise, the language on the vulnerability of computers came straight out of the Marsh Report, and the ideas on what to do about it were nearly identical to the plan that Clarke had written for Clinton.

The document set the framework for how cyber security would be handled over the next several years—as well as the limits in the government's ability to handle it at all, given industry's resistance to mandatory standards and (a problem that would soon become apparent) the Homeland Security Department's bureaucratic and technical inadequacies.

Clarke didn't stick around to fight the political battles of enforcing and refining the new plan. On March 19, Bush ordered the invasion of Iraq. In the buildup to the war, Clarke had argued that it would divert attention and resources from the fight against bin Laden and al Qaeda. Once the war's wheels were firmly in motion, Clarke resigned in protest.

But a few years after the invasion, as the war devolved from liberation to occupation and the enemy switched from Saddam Hussein to a disparate array of insurgents, the cyber warriors at Fort Meade and the Pentagon stepped onto the battlefield for the first time as a significant, even decisive force.

CHAPTER 9

CYBER WARS

W HEN General John Abizaid took the helm of U.S. Central Command on July 7, 2003, overseeing American military operations in the Middle East, Central Asia, and North Africa, his political bosses in Washington thought that the war in Iraq was over. After all, the Iraqi army had been routed, Saddam Hussein had fled, the Baathist regime had crumbled. But Abizaid knew that the war was just beginning, and he was flustered that President Bush and his top officials neither grasped its nature nor gave him the tools to fight it. One of those tools was cyber.

Abizaid had risen through the Army's ranks in airborne infantry, U.N. peacekeeping missions, and the upper echelon of Pentagon staff jobs. But early on in his career, he tasted a slice of the unconventional. In the mid-1980s, after serving as a company commander in the brief battle for Grenada, he was assigned to the Army Studies Group, which explored the future of combat. The Army vice chief of staff, General Max Thurman, was intrigued by reports of the Soviet army's research into remote sensing and psychic experiments. Nothing came of them, but they exposed Abizaid to the notion that war might be about more than just bullets and bombs.

In his next posting, as executive assistant to General John Shalikashvili, chairman of the Joint Chiefs of Staff, Abizaid once accompanied his boss on a trip to Moscow. Figuring their quarters were
bugged, the staff set up little tents so they could discuss official business away from Russian eavesdropping. Later, in Bosnia, as assistant
commander of the 1st Armored Division, Abizaid learned that the
CIA was flying unmanned reconnaissance planes over Sarajevo—
and he was aware of the worry, among U.S. intelligence officials
on the ground, that the Russians might seize control of a plane by
hacking its communications link.

By 2001, when Abizaid was promoted to director of the Joint Staff
in the Pentagon, the plans and programs for cyber security and cyber
warfare were in full bloom. His job placed him in the thick of squab-
bles and machinations among and within the services, so he knew
well the tensions between operators and spies throughout the cyber
realm. In the event of war, the operators, mainly in the military services, wanted to *use* the intelligence gleaned from cyber; the spies,
mainly in the NSA and CIA, saw the intelligence as vital for its own
sake and feared that using it would mean losing it—the enemy would
know that we'd been hacking into their networks, so they'd change
their codes or erect new barriers. Abizaid understood this tension—it
was a natural element in military politics—but he was, at heart, an
operator. He took the guided tour of Fort Meade, was impressed
with the wonders that the NSA could accomplish, and thought it
would be crazy to deny their fruits to American soldiers in battle.

In the lead-up to the invasion of Iraq, Abizaid, who was by now the
deputy head of Central Command, flew to Space Command headquarters in Colorado Springs, home of Joint Task Force-Computer
Network Operations, which would theoretically lead cyber offense
and defense in wartime. He was appalled by how bureaucratically
difficult it would be to muster any kind of cyber offensive campaign:
for one thing, the tools of cyber attack and cyber espionage were so

shrouded in secrecy that few military commanders even knew they existed.

Abizaid asked Major General James D. Bryan, the head of the joint task force, how he would go about getting intelligence from al Qaeda's computers into the hands of American soldiers in Afghanistan. Bryan traced the circuitous chain of command, from Space Command to a bevy of generals in the Pentagon, up to the deputy secretary of defense, then the secretary of defense, over to the National Security Council in the White House, and finally to the president. By the time the request cleared all these hurdles, the soldiers' need for the intel would probably have passed; the war itself might be over.

Bush ordered the invasion of Iraq on March 19. Three weeks later, after a remarkably swift armored assault up through the desert from Kuwait, Baghdad fell. On May Day, three weeks after the toppling, President Bush stood on the deck of the USS *Abraham Lincoln*, beneath a banner reading "Mission Accomplished," and declared that major combat operations were over. But later that month, the American proconsul, L. Paul Bremer, issued two directives, disbanding the Iraqi army and barring Baathist party members from power. The orders alienated the Sunni population so fiercely that, by the time Abizaid took over as CentCom commander, an insurgency was taking form, raging against both the new Shiite-led Iraqi government and its American protectors.

Abizaid heard about the vast reams of intelligence coming out of Iraq—communications intercepts, GPS data from insurgents' cell phones, photo imagery of Sunni jihadists flowing in from the Syrian border—but nobody was piecing the elements together, much less incorporating them into a military plan. Abizaid wanted to get inside those intercepts and send the insurgents false messages, directing them to a certain location, where U.S. special-ops forces would be lying in wait to kill them. But he needed cooperation from NSA and CIA to weave this intel together, and he needed authorization from political higher-ups to use it as an offensive tool. At the moment, he had neither.

The permanent bureaucracies at Langley and Fort Meade didn't want to cooperate: they knew that the world was watching—including the Russians and the Chinese—and they didn't want to waste their best intelligence-gathering techniques on a war that many of them regarded as less than vital. Meanwhile, Secretary of Defense Donald Rumsfeld wouldn't acknowledge that there *was* an insurgency. (Rumsfeld was old enough to know, from Vietnam days, that defeating an insurgency required a *counter*insurgency strategy, which in turn would leave tens of thousands of U.S. troops in Iraq for years, maybe decades—whereas he just wanted to get in, get out, and move on to oust the next tyrant standing in the way of America's post–Cold War dominance.)

Out of frustration, Abizaid turned to a one-star general named Keith Alexander. The two had graduated from West Point a year apart—Abizaid in the class of 1973, Alexander in '74—and they'd met again briefly, almost twenty years later, during battalion-command training in Italy. Now Alexander was in charge of the Army Intelligence and Security Command, at Fort Belvoir, Virginia, the land forces' own SIGINT center, with eleven thousand surveillance officers deployed worldwide—a mini-NSA all its own, but geared explicitly to Army missions. Maybe Alexander could help Abizaid put an operational slant on intelligence data.

He'd come to the right man. Alexander was something of a technical wizard. Back at West Point, he worked on computers in the electrical engineering and physics departments. In the early 1980s, at the Naval Postgraduate School, in Monterey, California, he built his own computer and developed a program that taught Army personnel how to make the transition from handwritten index cards to automated databases. Soon after graduating, he was assigned to the Army Intelligence Center, at Fort Huachuca, Arizona, where he spent his first weekend memorizing the technical specifications for all the Army's computers, then prepared a master plan for all intelligence and electronic-warfare data systems. In the run-up to Operation Desert Storm, the first Gulf War of 1991, Alexander led a team in the 1st Armored Division, at Fort Hood, Texas, wiring together a series of computers so that they could process data more efficiently. Rather than relying on printouts and manual indexing, the analysts and war planners back in the Pentagon could access data that was stored and sorted to their needs.

Before assuming his present command at Fort Belvoir, Alexander had been Central Command's chief intelligence officer. He told Abizaid about the spate of technical advances on the boards, most remarkably tools that could intercept signals from the chips in cell phones, either directly or through the switching nodes in the cellular network, allowing SIGINT teams to track the location and movements of Taliban fighters in Pakistan's northwest frontier or the insurgents in Iraq—even if their phones were turned off. This was a new weapon in the cyber arsenal; no one had yet exploited its possibilities, much less devised the procedures for one agency to share the intelligence feed with other agencies or with commanders in the field. Abizaid was keen to get this sharing process going.

Although CentCom oversaw American military operations in Iraq, Afghanistan, and their neighboring countries, its headquarters were in Tampa, Florida, so Abizaid made frequent trips to Washington. By August, one month into his tenure as its commander, intelligence on insurgents was flowing into Langley and Fort Meade. He could see the "ratlines" of foreign jihadists crossing into Iraq from Syria; he read transcripts of their phone conversations, which were correlated with maps of their precise locations. He wanted to give American soldiers access to this intel, so they could use it on the battlefield.

By this time, Keith Alexander had been promoted to the Army's deputy chief of staff for intelligence, inside the Pentagon, so he and Abizaid collaborated on the substantive issues and the bureaucratic politics. They found an ideal enabler in General Stanley McChrystal, head of the Joint Special Operations Command. If this new cache of intelligence made its way to the troops in the field, the shadow soldiers of JSOC would be the first troops to get and use it; and McChrystal, a soldier of spooky intensity, was keen to make that happen. All three worked their angles in the Pentagon and the intelligence community, but the main obstacle was Rumsfeld, who still refused to regard the Iraqi rebels as insurgents.

Finally, in January 2004, Abizaid arranged a meeting with President Bush and made the case for launching cyber offensive operations against the insurgents. Bush told his national security adviser, Condoleezza Rice, to put the subject on the agenda for the next NSC meeting. When it came up several days later, the deputies from the intelligence agencies knocked it down with the age-old argument: the intercepts were providing excellent information on the insurgents; *attacking* the source of the information would alert them (and other potential foes who might be watching) that they were being hacked, prompting them to change their codes or toss their cell phones, resulting in a major intelligence loss.

Meanwhile, the Iraqi insurgents were growing stronger, America was losing the war, and Bush was losing patience. Numbed by the resistance to new approaches and doubting that an outside army could make things right in Iraq anyway, Abizaid moved toward the view that, rather than redoubling its efforts, the United States should start getting out.

But then things started to change. Rumsfeld, disenchanted with all the top Army generals, passed over the standing candidates for the vacated post of Army chief of staff and, instead, summoned General Peter Schoomaker out of retirement.

Schoomaker had spent most of his career in Special Forces, an-

other smack in the face of regular Army. (General Norman Schwarzkopf, the hero of Desert Storm, had spoken for many of his peers
when he scoffed at Special Forces as out-of-control "snake eaters.")
McChrystal, who had long known and admired Schoomaker, told
him about the ideas that he, Abizaid, and Alexander had been trying
to push through. The new chief found them appealing but understood that they needed an advocate high up in the intelligence community. At the start of 2005, Mike Hayden was nearing the end of an
unusually long six-year tenure as director of the NSA. Schoomaker
urged Rumsfeld to replace him with Alexander.

Seventeen years had passed since an Army officer had run the
NSA; in its fifty-three-year history, just three of its directors had
been Army generals, compared with seven Air Force generals and
five Navy admirals. The pattern had reflected, and stiffened, the
agency's resistance to sharing intelligence with field commanders
of "small wars," who tended to be Army officers. Now the United
States was fighting a small war, which the sitting president considered a big deal; the Army, as usual, was taking the brunt of the ca-
sualties, and Alexander planned to use his new post to help turn the
fighting around.

McChrystal had already made breakthroughs in weaving together
the disparate strands of intelligence. He'd assumed command of
JSOC in September 2003. That same month, Rumsfeld signed an
executive order authorizing JSOC to take military action against al
Qaeda anywhere in the world without prior approval of the president
or notification of Congress. But McChrystal found himself unable to
do much with this infusion of great power: the Pentagon chiefs were
cut off from the combatant commands; the combatant commands
were cut off from the intelligence agencies. McChrystal saw al Qaeda
as a network, each cell's powers enhanced by its ties with other cells;
it would take a network to fight a network, and McChrystal set out
to build his own. He reached out to the CIA, the services' separate

intelligence bureaus, the National Geospatial-Intelligence Agency, the intel officers at CentCom. He prodded them into agreements to share data and imagery from satellites, drones, cell phone intercepts, and landline wiretaps. (When the Bush administration rebuilt the Iraqi phone system after Saddam's ouster, the CIA and NSA were let in to attach some devices.) But to make this happen—to fuse all this information into a coherent database and to transform it into an offensive weapon—he also needed the analytical tools and surveillance technology of the NSA.

That's where Alexander came in.

————

As Keith Alexander took over Fort Meade, on August 1, 2005, his predecessor, Mike Hayden, stepped down, seething with suspicion.

A few years earlier, when Alexander was running the Army Intelligence and Security Command at Fort Belvoir, the two men had clashed in a dragged-out struggle for turf and power, leaving Hayden with a bitter taste, a shudder of distrust, about every aspect and activity of the new man in charge.

From the moment Alexander assumed command at Fort Belvoir, he was determined to transform the place from an administrative center—narrowly charged with providing signals intelligence to Army units, subordinate to both the Army chief of staff and the NSA director—into a peer command, engaged in operations, specifically in the war on terror.

In his earlier post as CentCom's intelligence chief, Alexander had helped develop new analytic tools that processed massive quantities of data and parsed them for patterns and connections. He thought the technique—tracing telephone and email links (A was talking to B, who was talking to C, and on and on)—could help track down terrorists and unravel their networks. And it could serve as Alexander's entrée to the intelligence world's upper echelon.

But he needed to feed his software with data—and the place that had the data was the NSA. He asked Hayden to share it; Hayden turned him down. The databases were the agency's crown jewels, the product of decades of investments in collection technology, computers, and human capital. But Hayden's resistance wasn't just a matter of turf protection. For years, other rival intelligence agencies had sought access to Fort Meade's databases, in order to run some experiment or pursue an agenda of their own. But SIGINT analysis was an esoteric specialty; raw data could sire erroneous, even dangerous, conclusions if placed in untrained hands. And what Alexander wanted to do with the data—"traffic analysis," as NSA hands called it—was particularly prone to this tendency. Coincidences weren't proof of causation; a shared point of contact—say, a phone number that a few suspicious people happened to call—wasn't proof of a network, much less a conspiracy.

Fort Belvoir had a particularly flaky record of pushing precisely these sorts of flimsy connections. In 1999, two years before Alexander arrived, his predecessor, Major General Robert Noonan, had set up a special office called the Land Information Warfare Activity, soon changed to the Information Dominance Center. One of its experiments was to see whether a computer program could automatically detect patterns in data on the Internet—specifically, patterns indicating foreign penetration into American research and development programs.

Art Money, the assistant secretary of defense for command, control, communications, and intelligence, had funded the experiment, and, when it was finished, he and John Hamre, the deputy secretary of defense, went to Belvoir for a briefing. Noonan displayed a vast scroll of images and charts, showing President Clinton, former secretary of defense William Perry, and Microsoft CEO Bill Gates posing with Chinese officials: the inference seemed to be that China had infiltrated the highest ranks of American government and industry.

Hamre was outraged, especially since the briefing had already been shown to a few Republicans in Congress. Noonan tried to defend the program, saying that it wasn't meant as an intelligence analysis but rather as a sort of science-fair project, showing the technology's possibilities. Hamre wasn't amused; he shut the project down.

The architect of the project was Belvoir's chief technology adviser, a civilian engineer named James Heath. Intense, self-confident, and extremely introverted (when he talked with colleagues, he didn't look down at their shoes, he looked down at *his own* shoes), Heath was fanatical about the potential of tracking connections in big data—specifically what would later be called "metadata."

Hamre's slam might have meant the end of *some* careers, but Heath stayed on and, when Alexander took command of Fort Belvoir in early 2001, his fortunes revived. The two had known each other since the mid-1990s, when Alexander commanded the 525th Military Intelligence Brigade at Fort Bragg, North Carolina, and Heath was his science adviser. They were working on "data visualization" software even then, and Alexander was impressed with Heath's acumen and single-mindedness. Heath's workmates, even the friendly ones, referred to him as Alexander's "mad scientist."

One of Mike Hayden's concerns about Alexander's request for raw NSA data was that Heath would be the one running the data. This was another reason why Hayden denied the request.

But Alexander fought back. Soft-spoken, charming, even humorous in an awkward way that cloaked his aggressive ambition, he mounted a major lobbying campaign to get the data. He told anyone and everyone with any power or influence, especially on Capitol Hill and in the Pentagon, that he and his team at Fort Belvoir had developed powerful software for tracking down terrorists in a transformative way but that Michael Hayden was blocking progress and withholding data for parochial reasons.

Of course, Hayden had his own contacts, and he started to hear reports of this Army two-star's machinations. One of his sources even told him that Alexander was knocking on doors at the Justice Department, asking about the ways of the Foreign Intelligence Surveillance Court, which authorized warrants for intercepts of suspected agents and spies inside U.S. borders. This was NSA territory, and no one else had any business—legally, politically, or otherwise—sniffing around it.

Hayden started referring to Alexander as "the Nike swoosh," after the sneaker brand's logo (a fleet, curved line), which carried the slogan "Just do it"—a fitting summary, he thought, of Alexander's MO.

But Alexander won over Rumsfeld, who didn't much like Hayden and was well disposed to the argument that the NSA was too slow. Hayden read the handwriting on the wall and, in June 2001, worked out an arrangement to share certain databases with Fort Belvoir. The mutual distrust persisted: Alexander suspected that Hayden wasn't giving him all the good data; Hayden suspected that Alexander wasn't stripping the data of personal information about Americans who would unavoidably get caught up in the surveillance, as the law required.* In the end, the analytical tools that Alexander and Heath had so touted neither turned up new angles nor unveiled any terrorists. Hayden and Alexander both failed to detect signs of the September 11 attack.

Now, four years after 9/11, following a brief term as the Army's top intelligence officer in the Pentagon, Alexander was taking over the palace at Fort Meade, taking possession of the databases—and bringing along Heath as his scientific adviser.

---

*Ironically, while complaining that Alexander might not handle NSA data in a strictly legal manner, Hayden was carrying out a legally dubious domestic-surveillance program that mined the same NSA database, including phone conversations and Internet activity of American citizens. Hayden rationalized this program, code-named Stellar Wind, as proper because it had been ordered by President Bush and deemed lawful by Justice Department lawyers.

In his opening months on the job, Alexander had no time to push ahead with his metadata agenda. The top priority was the war in Iraq, which, for him, meant loosening the traditional strictures on NSA assets, putting SIGINT teams in regular contact with commanders on the ground, and tasking TAO—the elite hackers in the Office of Tailored Access Operations—to address the specific, the *tailored*, needs of General McChrystal's Special Forces in their fight against the insurgents.

He also had to repair some damage within NSA.

One week before Alexander arrived at Fort Meade, William Black, Hayden's deputy for the previous five years, pulled the plug on Trailblazer, the agency's gargantuan outsourced project to monitor, intercept, and sift communications from the digital global network.

Trailblazer had consumed $1.2 billion of the agency's budget since the start of the decade, and it had proved to be a disaster: a fount of corporate mismanagement, cost overruns, and—more to the point, as Alexander saw it—conceptual wrongheadedness. It was a monolithic system, built around massive computers to capture and process the deluge of digital data. The problem was that the design was too simple. Mathematical brute force worked in the era of analog signals intelligence, when an entire conversation or fax transmission spilled through the same wire or radio burst; but digital data streamed through cyberspace in packets, breaking up into tiny pieces, each of which traveled the fastest possible route before reassembling at the intended destination. It was no longer enough to collect signals from sensors out in the field, then process the data at headquarters: there were too many signals, racing too quickly through too many servers and networks. Trailblazer could be "scaled up" only so far, before the oceans of data overwhelmed it. Sensors had to process the information, and integrate it with the feed from other sensors, in *real time*.

Alexander's first task, then, was to replace Trailblazer—in other words, to devise a whole new approach to SIGINT for the digital age. His predecessors of the last decade had faced the same challenge, though less urgently. Ken Minihan possessed the vision, but lacked the managerial skills; Mike Hayden had the managerial acumen, but succumbed to the presumed expertise of outside contractors, who led him down a costly path to nowhere. Alexander was the first NSA director who understood the *technology* at the center of the enterprise, who could talk with the SIGINT operators, TAO hackers, and Information Assurance analysts on their own level. He was, at heart, one of them: more a computer geek than a policy maven. He would spend hours down on the floor with his fellow geeks, discussing the problems, the possible approaches, the solutions—so much so that his top aides installed more computers in his office on the building's eighth deck, so he could work on his beloved technical puzzles without spending too much time away from the broader issues and agendas that he needed to address as director.

As a result of his technical prowess and his ability to speak a common language with the technical personnel, he and his staff devised the conceptual outlines of a new system in a matter of months and launched the first stages of a new program within a year. They called it Turbulence.

Instead of a single, monolithic system that tried to do everything, Turbulence consisted of nine smaller systems. In part, the various systems served as backups or alternative approaches, in case the others failed or the global technology shifted. More to the point, each of the systems sliced into the network from a different angle. Some pieces intercepted signals from satellites, microwave, and cable communications; others went after cell phones; still others tapped into the Internet—and they went after Internet traffic on the level of data *packets*, the basic unit of the Internet itself, either tracking the packets from their origins or sitting on the backbone of Internet traffic (often with the cooperation of the major Internet service providers), detecting a target's packet, then alerting the hackers at TAO to take over.

It wasn't just Alexander's technical acumen that made Turbulence possible; it was also the huge advances—in data processing, storage, and indexing—that had taken place in just the previous few years. Alexander took over Fort Meade at just the moment when, in the world of computers, his desires converged with reality.

Over the ensuing decade, as Turbulence matured and splintered into specialized programs (with names like Turbine, Turmoil, QuantumTheory, QuantumInsert, and XKeyscore), it evolved into a thoroughly interconnected, truly global system that would make earlier generations of signals intelligence seem clunky by comparison.

Turbulence drew on the same massive databases as Trailblazer; what differed was the processing and sifting of the data, which were far more precise, more tailored to the search for specific information, and more closely shaped to the actual pathways—the packets and streams—of modern digital communications. And because the intercepts took place within the network, the target could be tracked on the spot, in real time.

In the early stages of Turbulence, a parallel program took off, derived from the same technical concepts, involving some of the same technical staff, but focused on a specific geographical region. It was called the RTRG—for Real Time Regional Gateway—and its first mission was to hunt down insurgents in Iraq.

RTRG got under way early in 2007, around the same time that General David Petraeus assumed command of U.S. forces in Iraq and President Bush ordered a "surge" in the number of those forces. Petraeus and Alexander had been friendly for more than thirty years: they'd been classmates at West Point, a source of bonding among Army officers, and they'd renewed their ties years later as brigade commanders at Fort Bragg. When they met again, as Petraeus led the fight in Baghdad, they made a natural team: Petraeus wanted to win the war through a revival of counterinsurgency techniques, and Alexander was keen to plow NSA resources into helping him.

Roadside bombs were the biggest threat to American soldiers in Iraq. Intelligence on the bombers and their locations flooded into NSA computers, from cell phone intercepts, drone and satellite imagery, and myriad other sources. But it took *sixteen hours* for the data to flow to the Pentagon, then to Fort Meade, then to the tech teams for analysis, then back to the intel centers in Baghdad, then to the soldiers in the field—and that was too long: the insurgents had already moved elsewhere.

Alexander proposed cutting out the middlemen and putting NSA equipment and analysts inside Iraq. Petraeus agreed. They first set up shop, a mini-NSA, in a heavily guarded concrete hangar at Balad Air Base, north of Baghdad. After a while, some of the analysts went out on patrol with the troops, collecting and processing data as they moved. Over the next few years, six thousand NSA officials were deployed to Iraq and, later, Afghanistan; twenty-two of them were killed, many of them by roadside bombs while they were out with the troops.

But their efforts had an impact: in the first few months, the lag time between collecting and acting on intelligence was slashed from sixteen hours to *one minute*.

By April, Special Forces were using this cache of intelligence to capture not only insurgents but also their computers; and stored inside those computers were emails, phone numbers, usernames, passwords of other insurgents, including al Qaeda leaders—the stuff of a modern spymaster's dreams.

Finally, Alexander and McChrystal had the ingredients for the cyber offensive campaign that they'd discussed with John Abizaid four years earlier. The NSA teams at Balad Air Base hoisted their full retinue of tricks and tradecraft. They intercepted insurgents' emails: in some cases, they merely monitored the exchanges to gain new intelligence; in other cases, they injected malware to shut down insurgents' servers; and in other—many other—cases, they sent phony emails to insurgents, ordering them to meet at a certain time, at a certain location, where U.S. Special Forces would be hiding and waiting to kill them.

In 2007 alone, these sorts of operations, enabled and assisted by the NSA, killed nearly four thousand Iraqi insurgents.

The effect was not decisive, nor was it meant to be: the idea was to provide some breathing space, a zone of security, for Iraq's political factions to settle their quarrels and form a unified state without having to worry about bombs blowing up every day. The problem was that the ruling faction, the Shiite government of Prime Minister Nouri al-Maliki, didn't want to settle its quarrels with rival factions among the Sunnis or Kurds; and so, after the American troops left, the sectarian fighting resumed.

But that pivotal year of 2007 saw a dramatic quelling of violence and the taming, co-optation, or surrender of nearly all the active militias. Petraeus's counterinsurgency strategy had something to do with this, as did Bush's troop surge. But the tactical gains could not have been won without the Real Time Regional Gateway of the NSA.

––––––––

RTRG wasn't the only innovation that the year saw in cyber offensive warfare.

On September 6, just past midnight, four Israeli F-15 fighter jets flew over an unfinished nuclear reactor in eastern Syria, which was being built with the help of North Korean scientists, and demolished it with a barrage of laser-guided bombs and missiles. Syrian president Bashar al-Assad was so stunned that he issued no public protest: better to pretend nothing happened than to acknowledge such a successful incursion. The Israelis said nothing either.

Assad was baffled. The previous February, his generals had installed new Russian air-defense batteries; the crews had been training ever since, and, owing to tensions on the Golan Heights, they'd been on duty the night of the attack; yet they reported seeing no planes on their radar screens.

The Israelis managed to pull off the attack—code-named Operation Orchard—because, ahead of time, Unit 8200, their secret cyber warfare bureau, had hacked the Syrian air-defense radar system. They did so with a computer program called Suter, developed by a clandestine U.S. Air Force bureau called Big Safari. Suter didn't disable the radar; instead, it disrupted the data link connecting the radar with the screens of the radar operators. At the same time, Suter hacked into the screens' video signal, so that the Unit 8200 crew could see what the radar operators were seeing. If all was going well, they would see blank screens—and all went well.

It harked back to the campaign waged in the Balkans, ten years earlier, when the Pentagon's J-39 unit, the NSA, and the CIA's Information Operations Center spoofed the Serbian air-defense command by tapping into its communications lines and sending false data to its radar screens. And the Serbian campaign had its roots in the plan dreamed up, five years earlier, by Ken Minihan's demondialers at the Air Force Information Warfare Center in San Antonio, to achieve air surprise in the (ultimately aborted) invasion of Haiti by jamming all the island's telephones.

The Serbian and Haitian campaigns were classic cases of information warfare in the pre-digital age, when the armed forces of many nations ran communications through commercial phone lines. Operation Orchard, like the NSA-JSOC operation in Iraq, exploited the growing reliance on computer networks. Haiti and the Balkans were experiments in *proto*-cyber warfare; Operation Orchard and the roundup of jihadists in Iraq marked the start of the real thing.

Four and a half months earlier, on April 27, 2007, riots broke out
in Tallinn, the capital of Estonia, the smallest and most Western-
leaning of the three former Soviet republics on the Baltic Sea, just
south of Finland. Estonians had chafed under Moscow's rule since
the beginning of World War II, when the occupation began. When
Mikhail Gorbachev took over the Kremlin and loosened his grip almost a half century later, Estonians led the region-wide rebellion for
independence that helped usher in the collapse of the Soviet Union.
When Vladimir Putin ascended to power at the turn of the twenty-
first century on a wave of resentment and nostalgia for the days of
great power, tensions once again sharpened.

The riots began when Estonia's president, under pressure from
Putin, overruled a law that would have removed all the monuments
that had gone up during the years of Soviet occupation, including a
giant bronze statue of a Red Army soldier. Thousands of Estonians
took to the streets in protest, rushing the bronze statue, trying to topple it themselves, only to be met by the town's ethnic Russians, who
fought back, seeing the protest as an insult to the motherland's wartime sacrifices. Police intervened and moved the statue elsewhere,
but street fights continued, at which point Putin intervened—not
with troops, as his predecessors might have done, but with an onslaught of ones and zeros.

The 1.3 million citizens of Estonia were among the most digitally
advanced on earth, a larger percentage of them hooked up to the Internet and were more reliant on broadband services than those of any
other country. The day after the Bronze Night riot, as it was called,
they were hit with a massive cyber attack, their networks and servers
flooded with so much data that they shut down. And unlike most
denial-of-service attacks, which tended to be one-off bits of mischief,
this attack persisted and was followed up—in three separate waves—

with infections of malware that spread from one computer to another, across the tiny nation, in all spheres of life. For three weeks, sporadically for a whole month, many Estonians were unable to use not just their computers but their telephones, bank accounts, credit cards: everything was hooked up to one network or another—the parliament, the government ministries, mass media, shops, public records, military communications—and it all broke down.

As a member of NATO, Estonia requested aid under Article 5 of the North Atlantic Treaty, which pledged each member-state to treat an attack on one as an attack on all. But the allies were skeptical. Was this an *attack*, in that sense? Was it an act of war? The question was left open. No troops were sent.

Nonetheless, Western computer specialists rushed to Estonia's defense at their own initiative, joining and aiding the considerable, skilled white-hat hacker movement inside Estonia. Using a variety of time-honored techniques, they tracked and expelled many of the intruders, softening the effects that would have erupted had the Tallinn government been the only source of resistance and defense.

Kremlin officials denied involvement in the attack, and the Westerners could find no *conclusive* evidence pointing to a single culprit—one reason, among several, for their reluctance to regard the cyber attacks as cause to invoke Article 5. Attributing the source of a cyber attack was an inherently difficult matter, and whoever launched this one had covered his tracks expertly. Still, forensic analysts did trace the malware code to a Cyrillic keyboard; in response, Kremlin authorities arrested a single member of the nationalist youth organization Nashi (the Russian word for "ours"), fined him the equivalent of a thousand dollars, and pronounced the crime solved. But no one believed that a single lowly citizen, or a small private group, could have found, much less hacked, some of the sensitive Estonian sites that had been taken down all at once and for such a long time.

———

The cyber strikes in Estonia proved to be the dress rehearsal for a coordinated military campaign, a little over a year later, in which Russia launched simultaneous air, ground, naval, and cyber operations against the former Soviet republic of Georgia.

Since the end of the Cold War, tensions had been rife between Moscow and the newly independent Georgian government over the tiny oblasts of South Ossetia and Abkhazia, formally a part of Georgia but dense with ethnic Russians. On August 1, 2008, Ossetian separatists shelled the Georgian village of Tskhinvali. The night of August 7–8, Georgian soldiers mobilized, suppressing the separatists and recapturing the town in a few hours. The next day, under the pretense of "peace enforcement," Russian troops and tanks rolled into the village, supported by air strikes and a naval blockade along the coast.

At the precise moment when the tanks and planes crossed the South Ossetian line, fifty-four Georgian websites—related to mass media, finance, government ministries, police, and armed forces— were hacked and, along with the nation's entire Internet service, rerouted to Russian servers, which shut them down. Georgian citizens couldn't gain access to information about what was happening; Georgian officers had trouble sending orders to their troops; Georgian politicians met long delays when trying to communicate with the rest of the world. As a result, Russian propaganda channels were first to beam Moscow's version of events to the world. It was a classic case of what was once called information warfare or counter command-control warfare—a campaign to confuse, bewilder, or disorient the enemy and thus weaken, delay, or destroy his ability to respond to a military attack.

The hackers also stole material from some sites that gave them valuable intelligence on the Georgian military—its operations, movements, and communiqués—so the Russian troops could roll over them all the more swiftly.

Just as with Estonia, Kremlin spokesmen denied launching the cyber attacks, though the timing—coordinated so precisely with the other forms of attack—splashed extreme doubt on their claims of innocence.

After four days of fighting, the Georgian army retreated. Soon after, Russia's parliament formally recognized South Ossetia and Abkhazia as independent states. Georgia and much of the rest of the world disputed the status, seeing the enclaves as occupied Georgian territory, but there wasn't much they could do about it.

In the sixteen months from April 2007 to August 2008, when America hacked Iraqi insurgents' email, Israel spoofed Syrian air defenses, and Russia flooded the servers of Estonia and Georgia, the world witnessed the dawn of a new era in cyber warfare—the fulfillment of a decade's worth of studies, simulations, and, at the start of the decade in Serbia, a tentative tryout.

The Estonian operation was a stab at political coercion, though in that sense it failed: in the end, the statue of the Red Army soldier was moved from the center of Tallinn to a military graveyard on the town's outskirts.

The other three operations were successes, but cyber's role in each was tactical: an adjunct to conventional military operations, in much the same way as radar, stealth technology, and electronic countermeasures had been in previous conflicts. Its effects were probably short-lived as well; had the conflicts gone on longer, the target-nations would likely have found ways to deflect, diffuse, or disable the cyber attacks, just as the Estonians did with the help of Western allies. Even in its four-day war in South Ossetia, Georgia managed to reroute some of its servers to Western countries and filter some of the Russian intrusions; the cyber attack evolved into a *two-way* cyber war, with improvised tactics and maneuvers.

In all its incarnations through the centuries, information warfare had been a gamble, its payoff lasting a brief spell, at best—just long enough for spies, troops, ships, or planes to cross a border unde-tected, or for a crucial message to be blocked or sent and received.

One question that remained about this latest chapter, in the In-ternet era, was whether ones and zeroes, zipping through cyberspace from half a world away, could inflict *physical* damage on a country's assets. The most alarming passages of the Marsh Report and a dozen other studies had pointed to the vulnerability of electrical power grids, oil and gas pipelines, dams, railroads, waterworks, and other pieces of a nation's critical infrastructure—all of them increasingly controlled by computers run on commercial systems. The studies warned that foreign intelligence agents, organized crime gangs, or malicious anarchists could take down these systems with cyber at-tacks from anywhere on earth. Some classified exercises, including the simulated phase of Eligible Receiver, posited such attacks. But were the scenarios plausible? Could a clever hacker *really* destroy a physical object?

On March 4, 2007, the Department of Energy conducted an ex-periment—called the Aurora Generator Test—to answer that ques-tion.

The test was run by a retired naval intelligence officer named Mi-chael Assante. Shortly after the 9/11 attacks, Assante was tasked to the FBI's National Infrastructure Protection Center, which had been set up in the wake of Solar Sunrise and Moonlight Maze, the first major cyber intrusions into American military networks. While most of the center's analysts focused on Internet viruses, Assante examined the vulnerability of the automated control systems that ran power grids, pipelines, and other pieces of critical infrastructure that the Marsh Report had catalogued.

A few years later, Assante retired from the Navy and went to work as vice president and chief security officer of the American Electrical Power Company, which delivered electricity to millions of customers throughout the South, Midwest, and Mid-Atlantic. Several times he raised these problems with his fellow executives. In response, they'd acknowledge that someone could hack into a control system and cause power outages, but, they would add, the damage would be short-term: a technician would replace the circuit breaker, and the lights would go back on. But Assante would shake his head. Back at the FBI, he'd talked with protection and control engineers, the specialists' specialists, who reminded him that circuit breakers were like fuses: their function was to protect very costly components, such as power generators, which were much harder, and would take much longer, to replace. A malicious hacker wouldn't likely stop at blowing the circuit breaker; he'd go on to damage or destroy the generator.

Finally persuaded that this might be a problem, Assante's bosses sent him to the Idaho National Laboratory, an 890-square-mile federal research facility in the prairie desert outside Idaho Falls, to examine the issues more deeply. First, he did mathematical analyses, then bench tests of miniaturized models, and finally set up a real-life experiment. The Department of Homeland Security had recently undertaken a project on the most worrisome dangers in cyberspace, so its managers agreed to help fund it.

The object of the Aurora test was a 2.25-megawatt power generator, weighing twenty-seven tons, installed inside one of the lab's test chambers. On a signal from Washington, where officials were watching the test on a video monitor, a technician typed a mere twenty-one lines of malicious code into a digital relay, which was wired to the generator. The code opened a circuit breaker in the generator's protection system, then closed it just before the system responded, throwing its operations out of sync. Almost instantly, the generator shook, and some parts flew off. A few seconds later, it shook again, then belched out a puff of white smoke and a huge cloud of black smoke. The machine was dead.

Before the test, Assante and his team figured that there would be damage; that's what their analyses and simulations had predicted. But they didn't expect the magnitude of damage or how quickly it would come. Start-up to breakdown, the test lasted just three minutes, and it would have lasted a minute or two shorter, except that the crews paused to assess each phase of damage before moving on.

If the military clashes of 2007—in Iraq, Syria, and the former Soviet republics—confirmed that cyber weapons could play a tactical role in new-age warfare, the Aurora Generator Test revealed that they might play a strategic role, too, as instruments of leverage or weapons of mass destruction, not unlike that of nuclear weapons. They would, of course, wreak much less destruction than atomic or hydrogen bombs, but they were much more accessible—no Manhattan Project was necessary, only the purchase of computers and the training of hackers—and their effects were lightning fast.

There had been similar, if less dramatic, demonstrations of these effects in the past. In 2000, a disgruntled former worker at an Australian water-treatment center hacked into its central computers and sent commands that disabled the pumps, allowing raw sewage to flow into the water. The following year, hackers broke into the servers of a California company that transmitted electrical power throughout the state, then probed its network for two weeks before getting caught.

The problem, in other words, was long known to be real, not just theoretical, but few companies had taken any steps to solve it. Nor had government agencies stepped in: those with the ability lacked the legal authority, while those with the legal authority lacked the ability; since action was difficult, evasion was easy. But for anyone who watched the video of the Aurora Generator Test, evasion was no longer an option.

One of the video's most interested viewers, who showed it to officials all around the capital, from the president on down, was the former NSA director who coined the phrase "information warfare," Rear Admiral Mike McConnell.

CHAPTER 10

BUCKSHOT YANKEE

O N February 20, two weeks before the Aurora Generator Test, Mike McConnell was sworn in as director of national intelligence. It was a new job in Washington, having been created just two years earlier, in the wake of the report by the 9/11 Commission concluding that al Qaeda's plot to attack the World Trade Center succeeded because the nation's scattered intelligence agencies—FBI, CIA, NSA, and the rest—didn't communicate with one another and so couldn't connect all the dots of data. The DNI, a cabinet-level post carrying the additional title of special adviser to the president, was envisioned as a sort of supra-director who would coordinate the activities and findings of the entire intelligence community; but many saw it as just another bureaucratic layer. When the position was created, President Bush offered it to Robert Gates, who had been CIA director and deputy national security adviser during his father's presidency, but Gates turned it down upon learning that he would have no power to set budgets or hire and fire personnel.

McConnell had no problem with the job's bureaucratic limits.

He took it with one goal in mind: to put cyber, especially cyber security, on the president's agenda.

Back in the early- to mid-1990s, as NSA director, McConnell had gone through the same roller-coaster ride that many others at Fort Meade had experienced: a thrilled rush at the marvels that the agency's SIGINT teams could perform—followed by the realization that whatever we can do to our enemies, our enemies could soon do to us: a dread deepened, in the decade since, by America's growing reliance on vulnerable computer networks.

After McConnell left the NSA in early 1996, he was hired by Booz Allen, one of the oldest management consulting firms along the capital's suburban Beltway, and transformed it into a powerhouse contractor for the U.S. intelligence agencies—an R&D center for SIGINT and cyber security programs, as well as a haven of employment for senior NSA and CIA officials as they ferried back and forth between the public and private sectors.

Taking the DNI job, McConnell gave up a seven-figure salary, but he saw it as a singular opportunity to push his passions on cyber into policy. (Besides, the sacrifice was hardly long-term; after his two-year stint back in government, he returned to the firm.) In pursuit of this goal, he stayed as close as he could to the Oval Office, delivering the president's intelligence briefing at the start of each day. A canny bureaucratic player with a casual drawl masking his laser-beam intensity, McConnell also dropped in, at key moments, on the aides and cabinet secretaries who had an interest in cyber security policy, whether or not they realized it. These included not only the usual suspects at State, Defense, and the National Security Council staff, but also the Departments of Treasury, Energy, and Commerce, since banks, utilities, and other corporations were particularly prone to attack. To McConnell's dismay, but not surprise, few of these officials displayed the slightest awareness of the problem.

So, McConnell pulled a neat trick out of his bag. He would bring the cabinet secretary a copy of a memo. Here, McConnell would say, handing it over. You wrote this memo last week. The Chinese hacked it from your computer. We hacked it back from their computer.

That grabbed their attention. Suddenly officials who'd never heard of cyber started taking a keen interest in the subject; a few asked McConnell for a full-scale briefing. Slowly, quietly, he was building a high-level constituency for his plan of action.

In late April, President Bush received a request to authorize cyber offensive operations against the insurgents in Iraq. This was the plan that Generals Abizaid, Petraeus, McChrystal, and Alexander had honed for months—finally sent up the chain of command through the new secretary of defense, Robert Gates, who had returned to government just two months earlier than McConnell, replacing the ousted Donald Rumsfeld.

From his experiences at the NSA and Booz Allen, McConnell understood the nature and importance of this proposal. Clearly, there were huge gains to be had from getting inside the insurgents' networks, disrupting their communications, sending them false emails on where to go, then dispatching a special-ops unit to kill them when they got there. But there were also risks: inserting malware into insurgents' email might infect other servers in the area, including those of American armed forces and of Iraqi civilians who had no involvement in the conflict. It was a complex endeavor, so McConnell scheduled an hour with the president to explain its full dimensions.

It was still a rare thing for a president to be briefed on cyber offensive operations—there hadn't been many of them, at this point—and the proposal came at a crucial moment: a few months into Bush's troop surge and the shift to a new strategy, new commander, and new defense secretary. So McConnell's briefing, which took place on May 16, was attended by a large group of advisers: Vice President Cheney, Secretary Gates, Secretary of State Condoleezza Rice, National Security Adviser Stephen Hadley, the Joint Chiefs of Staff vice chairman Admiral Edmund Giambastiani (the chairman, General Peter Pace, was traveling), Treasury Secretary Henry Paulson, and General Keith Alexander, the NSA director, in case someone asked about technical details.

As it turned out, there was no need for discussion. Bush quickly got the idea, finding the upside enticing and the downside trivial. Ten minutes into McConnell's hour-long briefing, he cut it short and approved the plan.

The room turned quiet. What was McConnell going to say now? He hadn't planned on the prospect, but it seemed an ideal moment to make the pitch that he'd taken this job to deliver. He switched gears and revved up the spiel.

Mr. President, he began, we come to talk with you about cyber *offense* because we need your permission to carry out those operations. But we don't talk with you much about cyber *defense*.

Bush looked at McConnell quizzically. He'd been briefed on the subject before, most fully when Richard Clarke wrote his *National Strategy to Secure Cyberspace*, but that was four years earlier, and a lot of crises had erupted since; cyber had never been more than a sporadic blip on his radar screen.

McConnell swiftly recited the talking points from two decades of analyses—the vulnerability of computer systems, their growing use in all aspects of American life, the graphic illustration supplied by the Aurora Generator Test, which had taken place just two months earlier. Then he raised the stakes, stating his case in the most urgent terms he could muster: those nineteen terrorists who mounted the 9/11 attack—if they'd been *cyber* smart, McConnell said, if they'd hacked into the servers of one major bank in New York City and contaminated its files, they could have inflicted more economic damage than they'd done by taking down the Twin Towers.

Bush turned to Henry Paulson, his treasury secretary. "Is this true, Hank?" he asked.

McConnell had discussed this very point with Paulson in a private meeting a week earlier. "Yes, Mr. President," he replied from the back of the room. The banking system relied on confidence, which an attack of this sort could severely damage.

Bush was furious. He got up and walked around the room. McConnell had put him in a spot, spelling out a threat and describing it as greater than *the* threat weighing on his and every other American's mind for the past five and a half years—the threat of another 9/11. And he'd done this in front of his most senior security advisers. Bush couldn't just let it pass.

"McConnell," he said, "*you* raised this problem. You've got thirty days to solve it."

It was a tall order: thirty days to solve a problem that had been kicking around for forty years. But at least he'd seized the president's attention. It was during precisely such moments—rare in the annals of this history—that leaps of progress in policy had been plotted: Ronald Reagan's innocent question after watching *WarGames* ("could something like this really happen?") led to the first presidential directive on computer security; Bill Clinton's crisis mentality in the wake of the Oklahoma City bombing spurred the vast stream of studies, working groups, and, at last, real institutional changes that turned cyber security into a mainstream public issue. Now, McConnell hoped, Bush's pique might unleash the next new wave of change.

McConnell had been surveying the landscape since returning to government, and he was shocked how little progress had been made in the decade that he'd been out of public life. The Pentagon and the military services had plugged a lot of the holes in their networks, but—despite the commissions, simulations, congressional hearings, and even the presidential decrees that Dick Clarke had written for

Clinton and Bush—conditions elsewhere in government, and still more so in the private sector, were no different, no less vulnerable to cyber attacks.

The reasons for this rut were also the same: private companies didn't want to spend the money on cyber security, and they resisted all regulations to make them do so; meanwhile, federal agencies lacked the talent or resources to do the job, except for the NSA, which had neither the legal authority nor the desire.

Entities had been created during the most recent spate of interest, during Clarke's reign as cyber coordinator under Clinton and the first two years of Bush, most notably the interagency Cyber Council and the ISACs—Information Sharing and Analysis Centers—that paired government experts with the private owners of companies involved in critical infrastructure (finance, electrical power, transportation, and so forth). But most of those projects stalled after Clarke resigned four years earlier. Now, with Bush's marching orders in hand, McConnell set out to bulk up these entities or create new ones, this time backed by serious money.

McConnell delegated the task to an interagency cyber task force, run by one of his assistants, Melissa Hathaway, the former director of an information operations unit at Booz Allen, whom he'd brought with him to be his chief cyber aide at the National Intelligence Directorate.

Protecting the civilian side of government from cyber attacks was new terrain. Fifteen years earlier, when the military services began to confront the problem, the first step they took was to equip their computers with intrusion-detection systems. So, as a first step, Hathaway's task force calculated what it would take to detect intrusions of *civilian* networks. The requirements turned out to be massive. When the tech crew at Kelly Air Force Base started monitoring computer networks in the mid-1990s, all of the Air Force servers, across the nation, had about one hundred points of access to the Internet.

Now, the myriad agencies and departments of the entire federal government had 4,300 access points.

More than this, the job of securing these points was assigned, by statute, to the Department of Homeland Security, a mongrel organization slapped together from twenty-two agencies, once under the auspices of eight separate departments. The idea had been to take all the agencies with even the slightest responsibility for protecting the nation from terrorist attacks and to consolidate them into a single, strong cabinet department. But in fact, the move only dispersed power, overloading the department's secretary with a portfolio much too large for any one person to manage and burying once-vibrant organizations—such as the Pentagon's National Communications System, which ran the alert programs for attacks of all sorts, including cyber attacks—in the dunes of a remote bureaucracy. The department was remote physically as well as politically, its headquarters crammed into a small campus on Nebraska Avenue in far Northwest Washington, five miles from the White House—the same campus where the NSA had stuck its Information Security Directorate until the late 1960s, when it was moved to the airport annex a half hour's drive (somewhat closer than Nebraska Avenue's hour-long trek) from Fort Meade.

In 2004, its second year of operations, the Homeland Security Department, in an outgrowth of one of Dick Clarke's initiatives, put out a contract for a government-wide intrusion-detection system, called Einstein. But the task proved unwieldy: the largest supercomputer would have had a hard time monitoring the traffic in and out of four thousand entryways to the Internet, and federal agencies weren't *required* to install the system in any case.

This mismatch between goals and capabilities set the stage for the new program put in motion by McConnell and Hathaway, which they called the Comprehensive National Cybersecurity Initiative, or CNCI. It called for the creation of a supra-agency that would con-

solidate the government's scattered servers into a single "Federal Enterprise Network," set strict security standards, and whittle down the
points of entry to the Internet from over four thousand to just fifty.

That was the goal, anyway.

On January 9, 2008, eight months after McConnell's big briefing, Bush signed a national security presidential directive, NSPD-54,
which cited the dangers posed by America's cyber vulnerabilities—
taking much of its language from a decade of directives and studies—
and ordered Hathaway's plan into action as the remedy.

In the weeks leading up to the directive, McConnell stressed that
the plan would be expensive; Bush waved away the warning, saying
that he was willing to spend as much money as Franklin Roosevelt
had spent on the Manhattan Project. Along with the White House
budget office, McConnell drew up a five-year plan amounting to $18
billion. The congressional intelligence committees cut only a small
slice, leaving him with $17.3 billion.

Although the plan's mission was to protect the computer networks of mainly civilian agencies, the entire program—the multibillion-dollar budget, the text of NSPD-54, even the existence of
something called the Comprehensive National Cybersecurity Initiative—was stamped Top Secret. Like most matters cyber, it was
bound up with the blackout secrecy of the NSA, and this was no
coincidence: on paper, the Department of Homeland Security was
the initiative's lead agency, but the NSA was placed in charge of technical support; and since neither Homeland Security nor any other
agency had the know-how or resources to *do* what the president's
directive wanted done, the locus of power, for this program, too,
would tilt from the campus on Nebraska Avenue to the sprawling
complex at Fort Meade.

Keith Alexander, the director of NSA, was also more adept at
budget politics than the managers at Homeland Security. He knew,
as Mike Hayden had before him, which legal statutes authorized

which sets of activities (Title 50 for intelligence, Title 10 for military operations, Title 18 for criminal probes) and which congressional committees dished out the money for each. So, when the initiative's $17.3 billion was divvied up among the various agencies, the vast bulk of it went to NSA—which, after all, would be buying and maintaining the hardware, the program's costliest element. Congress specified that Fort Meade spend its share of the sum on cyber defense. But that term was loosely defined, and the NSA budget was highly classified, so Alexander allocated the funds as he saw fit.

Meanwhile, Homeland Security upgraded Einstein, the inadequate intrusion-detection system, to Einstein 2, which was designed not only to detect malicious activity on a network, but also to send out an automatic alert. And the department started drawing the conceptual blueprints for Einstein 3, which—again, in theory—would automatically repel intruders. The NSA took on these projects as part of its share of the $17.3 billion, integrating them with the massive data-gathering, data-crunching enterprises it had already launched. But soon after joining forces on the Einstein project, Alexander backed out, explaining that the civilian agencies' requirements and Homeland Security's approach were incompatible with NSA's. Einstein's commercial contractors stayed on, and Homeland Security hired a team of cyber specialists, but, left to themselves, they had to start over; the program bogged down, fell short of its goals, and went into a tailspin.

And so, despite the president's full commitment and heaps of money, the vulnerability of computers and its implications for national security, economic health, and social cohesion—a topic that had set off intermittent alarm bells through the previous four decades—drifted once again into neglect.

Alexander was still obligated to spend his share of the money on cyber defense, but by this time, Ken Minihan's epiphany—that cyber offense and cyber defense ran on the same technology, were practically synonymous—had been fully ingrained in Fort Meade thinking.

The basic concepts of cyber were still in circulation—Computer Network *Attack*, Computer Network *Defense*, and Computer Network *Exploitation*—but the wild card was, and always had been, *exploitation*, CNE: the art and science of finding and exploiting vulnerabilities in the adversary's network, getting inside it, and twisting it around. CNE could be seen, used, and justified as preparation for a future cyber attack *or* as a form of what strategists had long called "active defense": penetrating an adversary's network to see what kinds of attacks he was planning, so that the NSA could devise a way to disrupt, degrade, or defeat them preemptively.

Alexander put out the word that, as in other types of warfare, active defense was essential: some cyber equivalent of the Maginot Line or the Great Wall of China wouldn't hold in the long run; adversaries would find a way to maneuver around or leap over the barriers. So, in the interagency councils and behind-closed-doors testimony, Alexander made the case that his piece of the Comprehensive National Cybersecurity Initiative should focus on CNE. And of course, once the money was lavished on tools for CNE, they could be programmed for offense *and* defense, since CNE was an enabler of both. When Alexander penetrated and probed the email and cell phone networks of Iraqi insurgents, that was CNE; when President Bush authorized him to disable and disrupt those networks—to intercept and send false messages that wound up getting insurgents killed—that was CNA, Computer Network Attack. Except for the final step, the decision to attack, CNE and CNA were identical.

Regardless of anyone's intentions (and Alexander's intentions were clear), this was the nature of the technology—which made it all the more vital for *political* leaders to take firm control: to ensure that policy shaped the use of technology, not the other way around.

Yet, just as cyber tools were melding into weapons of war, and as computer networks were controlling nearly every facet of daily life, the power shifted subtly, then suddenly, to the technology's masters at Fort Meade.

———

The pivotal moment in this shift occurred at NSA headquarters on Friday, October 24, 2008. At two-thirty that afternoon, a team of SIGINT analysts noticed something strange going on in the networks of U.S. Central Command, the headquarters running the wars in Afghanistan and Iraq.

A beacon was emitting a signal, and it seemed to be coming from inside CentCom's *classified* computers. This was not only strange, it was supposedly impossible: the military's classified networks weren't connected to the public Internet; the two were separated by an "air gap," which, everyone said, couldn't be crossed by the wiliest hacker. And yet, somehow, someone had made the leap and injected a few lines of malicious code—that was the only plausible source of the beacon—into one of the military's most secure lines of communication.

It was the first time ever, as far as anyone knew, that a classified network of the Department of Defense had been hacked.

The intrusion might not have been spotted, except that, a year earlier, when cyber war took off as a worldwide phenomenon, Richard Schaeffer, head of the NSA's Information Assurance Directorate—whose staff spent their workdays mulling and testing new ways that an outsider might breach its defenses—dreamed up a new tangent. Over the previous decade, the military services and the various joint task forces had done a reasonably good job of protecting the perimeters of their networks. But what if they'd missed something and an adversary was already inside, burrowing, undetected, through thousands or millions of files, copying or corrupting their contents?

Schaeffer assigned his Red Team—the same unit that had run the Eligible Receiver exercise back in 1997—to scan the classified networks. This team discovered the beacon. It was attached to a worm that they'd seen a couple years earlier under the rubric agent.btz. It was an elegant device: after penetrating the network and scooping up data, the beacon was programmed to carry it all home. The Office of Tailored Access Operations, the NSA's cyber black-bag shop, had long ago devised a similar tool.

Schaeffer brought the news to Alexander. Within five minutes, the two men and their staffs came up with a solution. The beacon was programmed to go home; so, they said, let's get inside the beacon and reroute it to a different home—specifically, an NSA storage bin. The idea seemed promising. Alexander put his technical teams on the task. Within a few hours, they figured out how to design the software. By the following morning, they'd created the program. Then they tested it on a computer at Fort Meade, first injecting the agent.btz worm, then zapping it with the rerouting instruction. The test was a success.

It was two-thirty, Saturday afternoon. In just twenty-four hours, the NSA had invented, built, and verified a solution. They called the operation Buckshot Yankee.

Meanwhile, the analytical branches of the agency were tracing the worm's pathways back to its starting point. They speculated that a U.S. serviceman or woman in Afghanistan had bought a malware-infected thumb drive and inserted it into a secure computer. (A detailed analysis, over the next few months, confirmed this hypothesis.) Thumb drives were widely sold at kiosks in Kabul, including those near NATO's military headquarters. It turned out, Russia had supplied many of these thumb drives, some of them preprogrammed by an intelligence agency, in the hopes that, someday, some American would do what—it now seemed clear—some American had actually done.

But all that was detail. The big picture was that, on the Monday morning after the crisis began, Pentagon officials were scrambling to grasp the scope of the problem—while, two days earlier, the NSA had solved it.

Admiral Mike Mullen, chairman of the Joint Chiefs of Staff, called an emergency meeting Monday morning to discuss a course of action, only to find that the service chiefs had sent mere colonels to attend. "What are *you* doing here?" he almost hollered. The networks of the nation's active war command had been compromised; it couldn't win battles without confidence in those networks. He needed to talk with the commanders and with the Joint Staff's directors of operations and intelligence—that is to say, he needed to talk with three- and four-star generals and admirals.

Later that morning, Mullen arranged a teleconference call with Mike McConnell, Keith Alexander, and General Kevin Chilton, the head of U.S. Strategic Command, which housed Joint Task Force-Global Network Operations, the latest incarnation of the loosely structured bureaus that had first been set up, a decade earlier, as Joint Task Force-Computer Network Defense.

Mullen started off the call with the same question that John Hamre had asked back in 1998, in the wake of Solar Sunrise, the first deep penetration of military networks: *Who's in charge?*

For twenty-five years, ever since Ronald Reagan signed the first presidential directive on computer security, the White House, the Pentagon, Congress, Fort Meade, and the various information warfare centers of the military services had been quarreling over that question. Now, General Chilton insisted that, because Strategic Command housed JTF-GNO, he was in charge.

"Then what's the plan?" Mullen asked.

Chilton paused and said, "Tell him, Keith."

Clearly, StratCom had nothing. No entity, civilian or military, had anything—any ideas about who'd done this, how to stop it, and what to do next—except for the agency with most of the money, technology, and talent to deal with such questions: the NSA.

The NSA directors of the past decade had worked feverishly to keep the business at Fort Meade in the face of competition from the services' scattershot cyber bureaus—"preserving the mystique," as Bill Perry had described the mission to Ken Minihan. The best way to do this was to make the case, day by day, that NSA was the only place that knew how to do this sort of thing, and that's what Alexander dramatized with Buckshot Yankee.

Bob Gates watched over this contrast between Fort Meade's control and the Pentagon's scramble with a mixture of horror and bemusement. He had been secretary of defense for nearly two years, after a long career in the CIA and a brief spell in the White House of Bush's father, and he continued to marvel at the sheer dysfunction of the Pentagon bureaucracy. When he first took the job, the military was locked in the grip of two wars, both going badly, yet the building's vast array of senior officers acted as if the world was at peace: they were pushing the same gold-plated weapons, built for some mythic major war of the future, that they'd been pushing since the Cold War, and promoting the same kinds of salute-snapping, card-punching officers—in short, they were doing nothing of any use—until he fired a few generals and replaced them with officers who seemed able and willing to help the men and women fighting, dying, and getting hideously injured in the wars that were happening now.

Almost every day since coming to the Pentagon, Gates had heard briefings on the latest attempt, by some serious adversary or mischievous hacker, to penetrate the Defense Department's networks. Here was the really serious breach that many had warned might happen, and, still, everyone was playing bureaucratic games; nobody seemed to recognize the obvious.

Mike McConnell, who'd been friendly with Gates since his time as NSA director, had been repeatedly making the case for a unified Cyber Command, which would supersede all the scattered cyber bureaus, run offensive *and* defensive operations (since they involved the same technology, activities, and skills), and ideally be located at Fort Meade (since that was where the technology, activities, and skills were concentrated). McConnell backed up his argument with a piece of inside knowledge: the NSA didn't like to share intelligence with operational commands; the only way to get it to do so was to fuse the NSA director and the cyber commander into the same person.

Gates had long thought McConnell's idea made sense, and Buckshot Yankee drove the point home.

Another development laced this point with urgency. The clock was ticking on Alexander's tenure at NSA. Most directors had served a three-year term; Alexander had been there for three years and two months. Beyond the math, Gates had heard rumors that Alexander was planning to retire, not just from the agency but also from the Army. Gates thought this would be disastrous: the CIA had recently predicted a major cyber attack in the next two years; here we were, in a crisis of lesser but still serious magnitude, and Alexander was the only official with a grip on what was happening.

The NSA director, by custom, was a three-star general or admiral; the heads of military commands were four-stars. Gates figured that one way to consolidate cyber policy and keep Alexander onboard was to create a new Cyber Command, write its charter so that the commander would also be the NSA director (as McConnell had suggested), and put Alexander in the double-hatted position, thus giving him a fourth star—and at least another three years on the job.

In fact, the rumors of Alexander's imminent departure were untrue. By coincidence, not long before Buckshot Yankee, Alexander made an appointment for a retirement briefing that generals were required to receive upon earning a third star. Alexander had put off his session for months; these things were usually a waste of time, and he was busy. Finally, the Army personnel command applied pressure, so he went to the next scheduled briefing.

Two days later, he got a call from Gates, wanting to know if rumors of his retirement were true. Alexander assured him they were not. Nonetheless, Gates told him of the plan to get him a fourth star.

It would take several months to line up the pins in the Pentagon, the intelligence community, and the Congress. Meanwhile, an election took place, and a new president, Barack Obama, arrived at the White House. But Gates, who agreed to stay on as defense secretary for at least a year, pushed the idea through. On June 23, 2009, he signed a memorandum, ordering the creation of U.S. Cyber Command.

———

During the final year of Bush's presidency and the first few months of Obama's, Gates wrestled with a dilemma. He'd realized for some time that, when it came to cyber security, there was no substitute for Fort Meade. The idea of turning the Department of Homeland Security into an NSA for civilian infrastructure, a notion that some in the White House still harbored, was a pipe dream. DHS didn't have the money, the manpower, or the technical talent—and, realistically, it never would. Yet because NSA was legally (and properly) barred from domestic surveillance, *it* couldn't protect civilian infrastructure, either.

On July 7, 2010, Gates had lunch at the Pentagon with Janet Napolitano, the secretary of homeland security, to propose a way out of the thicket. The idea was this: she would appoint a second deputy director of the NSA (Gates would have to name the person formally, but it would be her pick); in the event of a threat to the nation's critical infrastructure, this new deputy could draw on the technical resources of the NSA while invoking the legal authority of DHS.

Napolitano liked the idea. At a subsequent meeting, they drew up a memorandum of understanding on this arrangement, which included a set of firewalls to protect privacy and civil liberties. General Alexander, whom they consulted, gave it his blessings. On July 27, less than three weeks after their initial lunch, Gates and Napolitano took the idea to President Obama. He had no objections and passed it on to Thomas Donilon, his national security adviser, who vetted the idea with an interagency panel of the National Security Council.

Everything seemed on course. Gates and Napolitano left the details to their underlings and went back to more urgent business.

Over the next few months, the arrangement unraveled.

Before delegating the matter, Napolitano selected her candidate for the cyber deputy director—a two-star admiral named Michael Brown, who was her department's deputy assistant secretary for cyber security. Brown seemed ideal for the job. He'd studied math and cryptology at the Naval Academy, worked on SIGINT teams at the NSA, and, in the late 1990s, moved over to the Pentagon as one of the charter analysts—dealing with the Solar Sunrise and Moonlight Maze hacks—at Joint Task Force-Computer Network Defense. When Mike McConnell convinced President Bush to spend $18 billion on cyber security, he asked Brown to go work at the Department of Homeland Security, to help protect civilian networks in the same way that he'd helped protect military networks. For the next two years, that's what Brown tried to do, expanding the DHS cyber staff from twenty-eight people to roughly four hundred and turning its computer emergency response team into a vaguely functional organization. If there was someone who could merge the cultures of NSA and DHS, it was likely to be Mike Brown.

For that reason, though, he ran into obstacles at every step. Napolitano's deputy, Jane Holl Lute—a lawyer, former assistant secretary-general for peacekeeping support at the United Nations, and an Army veteran in signals intelligence—was deeply suspicious of NSA and resistant to any plan that would give the agency any power in domestic matters or that might turn the Internet into a "war zone." The same was true of the White House cyber security adviser, Howard Schmidt, who winced at those who described cyberspace as a "domain," in the same sense that Air Force and Navy officers described the skies and oceans as "domains" for military operations. Brown's rank as a naval officer, his background in cryptology, and his experience with the NSA suggested that this joint endeavor would be far from an equal partnership—that Fort Meade would run the show.

There was also resistance among the department deputies in the National Security Council, some of whom were peeved that this deal had gone down without their consultation. In the end, they approved Brown as "cybersecurity coordinator," but they wouldn't let him be a deputy director of the NSA; they wouldn't give him the legal authority he'd need to do the job that Gates and Napolitano had envisioned.

It was reminiscent, though few remembered so far back, of the dispute more than a quarter century earlier, in 1984, when civil liberties advocates in Congress resisted the plan—laid out in President Reagan's directive, NSDD-145—to put standards for computer security in the hands of a committee run by the director of the NSA.

The staff meetings between DHS and NSA practically seethed with tension. The Gates-Napolitano plan called for each agency to send ten analysts to the other's headquarters as a sort of cultural exchange. Early on, Fort Meade sent its ten—nine from NSA, one from Cyber Command—but DHS was slow to reciprocate. Part of the problem was simple logistics. Twenty-five thousand people worked at NSA; trading ten of them required scant sacrifice. But

DHS had only a few hundred cyber specialists; rather than transferring any, Lute decided to hire ten new people, a process that involved juggling the budget, vetting security clearances—in short, time: lots of time. Well before all ten came onboard, the arrangement sputtered, its wheels grinding nearly to a halt.

On October 31, 2010, U.S. Cyber Command raised its flag at Fort Meade, with General Alexander at the helm while, simultaneously, entering his sixth year as director of an NSA that was teeming with unprecedented political, bureaucratic, and computational power.

CHAPTER 11

"THE WHOLE HAYSTACK"

URING the early weeks of Mike McConnell's tenure as director of national intelligence, in 2007, one of his assistants showed him a chart produced by VeriSign, the company that operated the domain name system, which registered dot-com, dot-gov, dot-net, and other email addresses that made the Internet function. The chart displayed a map of the globe, laid out not by the geography of landmass and oceans but rather by the patterns and densities of network bandwidth. According to this map, 80 percent of the world's digital communications passed through the United States.*

The implications for intelligence were profound. If a terrorist in Pakistan was exchanging email or talking on a cell phone with an

---

*In the late 1990s, when he started researching the vulnerability of infrastructure, Richard Clarke learned that 80 percent of global Internet traffic passed through just two *buildings* in the United States: one, called MAE West (MAE standing for Metropolitan Area Exchange), in San Jose, California; the other, MAE East, above a steakhouse in Tysons Corner, Virginia. One night, Clarke took a Secret Service agent to dinner at the steakhouse, after which they took a look at the room upstairs. (He brought the agent along to avoid getting arrested.) They were both shocked at how easily a saboteur could wreak devastating damage.

191

arms supplier in Syria, and if the global network routed a piece of their communication through the United States, there was no need to set up a data-scooping station in hostile territory; the NSA could simply tap into the stream stateside.

But there was a legal obstacle. Back in the 1970s, hearings chaired by Senator Frank Church uncovered massive abuse by the CIA and NSA, including surveillance of American citizens, mainly political critics and antiwar activists, in violation of their Fourth Amendment rights against "unreasonable searches and seizures." The hearings led to the passage, in 1978, of the Foreign Intelligence Surveillance Act, which barred domestic surveillance without evidence of probable cause that the target was an agent of a foreign power and that the places of surveillance were, or would be, used by that agent; and even then, the government would have to present the evidence of probable cause to a secret FISA court, whose judges would be appointed by the chief justice of the U.S. Supreme Court. The president could authorize surveillance without a court order, but only if the attorney general certified under oath that the target was believed to be a foreign agent *and* that the eavesdropping would not pick up the communications of a "United States person," defined as an American citizen, permanent resident, or corporation.

After the attacks of September 11, 2001, Congress hurriedly passed the Patriot Act, which, among other things, revised FISA to allow surveillance of not only foreign agents but also members of amorphous terrorist groups, such as al Qaeda, which had no affiliation with a nation-state.

To McConnell's mind, even with that revision, FISA was out of date and in need of change. In the digital age, there were no discrete *places* of surveillance; cyberspace was everywhere. Nor could the government honestly certify that, while intercepting a terrorist's email or cell phone conversation, it wouldn't also pick up some innocent American's chatter; this was the nature of data packets, which whooshed pieces of many communications through the most efficient path. Since the most efficient path often ran through the United States, it would be hard *not* to pick up some Americans' data in the process.

In briefings to the president, meetings with national security aides, and informal sessions with members of Congress, McConnell brought along the VeriSign map, explained its implications, and made his pitch to amend FISA.

He knew that he was making progress when he met with Jack Murtha, ranking Democrat on the House appropriations defense subcommittee. Seventy-four years old, in his seventeenth term in Congress, Murtha had given McConnell a hard time back in the 1970s when he was NSA director; at one point, Murtha threatened to kill the agency's information warfare programs, especially those that had an offensive tilt. But the VeriSign map riveted his attention.

"Look at where all the bandwidth is," McConnell said, pointing to the bulge on American territory. "We need to change the law to give us access." Murtha bought the pitch, and so did almost everyone else who heard it.

President Bush needed no special pleading. Keen to do anything that might catch a terrorist in the act, he found a rationale for action in the VeriSign map and told his legal team to draft a bill.

On July 28, in his Saturday radio address, Bush announced that he was sending the bill to Congress. In the age of cell phones and the Internet, he said, current laws were "badly out of date," and, as a result, "we are missing a significant amount of foreign intelligence that we should be collecting to protect our country."

Four days later, the Senate's Republican leaders brought the bill to the floor as the Protect America Act. It did everything McConnell wanted it to do. One key passage noted that "electronic surveillance" of an American would not be illegal—would not even be defined as "electronic surveillance"—if it was aimed at a person who was "reasonably believed to be located outside of the United States." The stray collection of Americans' data, unavoidable in the digital world, would thus be exempt from possible prosecution. Another clause clarified that, under this new law, the attorney general's certification to the FISA court "is not required to identify the specific facilities, places, premises, or property" where the intelligence gathering would take place. As McConnell had been saying over and over, targets of surveillance in the digital age—unlike those in the era of phone taps—occupied no physical space.

One other significant passage specified that the government could obtain this information only "with the assistance of a communications service provider." This provision was barely noticed; to the extent it was, it seemed like a restriction, but in fact it gave the NSA license to retrieve data from private corporations—and gave the corporations legal cover to cooperate with the NSA. Few outsiders knew that service providers—from Western Union and AT&T in the early days, to Sprint and Verizon in the "Baby Bells" era, to Microsoft, Google, and the other pioneers of the Internet age—had long enjoyed mutually beneficial arrangements with the NSA and FBI. This section of the bill would loom large, and incite enormous controversy, six years later, when Edward Snowden's leaks revealed the vast extent of those arrangements.

Except for the requirement to consult with the FISA Court and the select congressional committees, both of which met in secrecy, the only limit that the bill placed on surveillance was that the data acquired from Americans—whose communications often got swept up along with the packets of data under surveillance—had to be "minimized." This meant that, to preserve privacy and civil liberties, the government could not store the *names* of any Americans or the *contents* of their communications, but rather only their phone numbers and the date, time, and duration of a conversation. Few who read the bill understood the definition of "minimized" or grasped how much even this amount of information—metadata, it was called—could reveal about someone's identity and activities.

After two days of debate, the Senate passed the measure, 60–28. The next day, the House concurred, 227–183. The following day, August 5, 2007, just eight days after his radio address, President Bush signed the bill into law.

With the technical advances of the previous decade—the Turbulence program, the Real Time Regional Gateway, the new generation of supercomputers, and the ingenuity of the hackers in the Office of Tailored Access Operations—the government could wade into all the streams of the World Wide Web. And with the new political powers invested in Fort Meade—the consolidation of all the services' signals intelligence bureaus and the start-up of U.S. Cyber Command, to be headed by the NSA director—it would be the NSA that did the wading, with the consent and authority of the White House, the Congress, and the secret chamber that the Supreme Court had set up as its proxy in the dark world.

It was a new age of expansive horizons for the NSA, and Keith Alexander was its ideal voyager. A common critique of the intelligence failure on 9/11 was that the relevant agencies possessed a lot of facts—a lot of data points—that might have pointed to an imminent attack, but no one could "connect the dots." Now, six years later, new technology allowed the NSA to gather so much data—a seamless stream of data points—that the dots almost connected themselves.

The convergence of technological advances and the post-9/11 fears of terrorism spawned a cultural change, too: a growing, if somewhat resigned, acceptance of intrusions into daily life. Back in 1984, the first presidential directive on computer security, signed by Ronald Reagan, was quashed because it empowered the NSA to set standards for all American computers—military, government, private, and commercial—and Congress wasn't about to let Fort Meade have a say in *domestic* surveillance or policy. Now, not quite a quarter century later, digital data crossed borders routinely; for all practical purposes, borders withered away, and so did the geographic strictures on the reach of the NSA.

In this context, Alexander saw an opening to revive the metadata program that he'd created, back at the start of the decade, as head of the Army Intelligence and Security Command at Fort Belvoir. The case for a revival seemed a logical fit to the technical, political, and cultural trends. Let's say, Alexander would argue, that, while tracking foreign communications, SIGINT operators spotted an American phone number calling the number of a known terrorist in Pakistan. The NSA could seek a warrant from the FISA Court to find out more about this American. They might also find it useful to learn other phone numbers that the suspicious American had been calling, and then perhaps to track what numbers *those* people had been calling. And, just like the Belvoir experiment, but writ large, it wouldn't take long before the NSA had stored data on millions of people, many of them Americans with no real connection to terrorism.

Then came a modern twist. At some point, Alexander's argument continued, the SIGINT analysts might find some American engaged in genuinely suspicious activity. They might wish that they'd been tracking this person for months, even years, so they could search through the data for a pattern of threats, possibly a nexus of conspirators, and trace it back to its origins. Therefore, it made sense to scoop up and store *everything* from *everybody*. NSA lawyers even altered some otherwise plain definitions, so that doing this didn't constitute "collecting" data from American citizens, as that would be illegal: under the new terminology, the NSA was just *storing* the data; the *collecting* wouldn't happen until an analyst went to retrieve it from the files, and that would be done only with proper FISA Court approval.

Under the FISA law, data could be stored only if it was deemed "relevant" to an investigation of foreign intelligence or terrorism.

But under this new definition, *everything* was potentially relevant: there was no way of knowing what *was* relevant until it *became* relevant; therefore, you had to have everything on hand to make a definitive assessment. If much of intelligence involved finding a needle in a haystack, Alexander liked to say, you had to have access to "the whole haystack."

The FISA Court had been created to approve, deny, or modify specific requests for collection; in that sense, it was more like a municipal court than the Supreme Court. But in this instance, the FISA Court ruled on the NSA's broad interpretation of the law, and it endorsed this definition of "relevant."

Keith Alexander had the whole haystack.

This was the state of cyberspace—a web thoroughly combed, plowed, and penetrated by intelligence services worldwide, especially America's own—when Senators Barack Obama and John McCain wound down their race for the White House in the fall of 2008.

President Bush was obligated to provide intelligence briefings to both candidates, so, on September 12, he sent Mike McConnell to brief Obama, the Democrat, at his campaign headquarters in Chicago. Bush was leery of the whole exercise and told McConnell, before he left, not to divulge anything about operations in Afghanistan and Iraq—and to brief only the candidate, not anyone on his staff.

Two of Obama's staff members, who'd planned to sit in and take notes, were miffed when the intelligence director asked them to leave; so was the candidate. But the session proceeded in a cordial enough way, Obama saying that he didn't want to hear anything about Iraq or Afghanistan, he had his own ideas on those wars; what he really wanted to discuss was terrorism.

McConnell went through the threats posed by al Qaeda and its affiliates, the various plots, some only barely disrupted, at home and abroad. Obama, who'd been a junior member of the Senate Foreign Relations Committee but had never heard such a detailed briefing from such a high-level intelligence official, was fascinated. At this point, fifty minutes had passed, more time than his aides had scheduled, but Obama was settling in; he asked McConnell what else he had in his briefing book.

Glad to oblige, the intelligence director told the next president about the status of North Korea's plan to detonate an A-bomb, Iran's program to build one, and Syria's atomic reactor in the desert. (Israel had bombed it a year earlier, but Assad was still in touch with his Pyongyang suppliers.) This took up another twenty minutes. Obama told him to go on.

And so, just as he'd done when Bush approved the Iraq cyber offensive plan after ten minutes of a meeting that had been scheduled for an hour, McConnell turned to the topic of his deepest worries. Earlier in the year, U.S. officials had alerted both Obama and McCain that China had hacked into their campaign computer systems, rummaging through their position papers, finances, and emails.

"They *exploited* your system," McConnell said. "What if they'd *destroyed* it?"

"That would have been problematic for me," Obama replied.

"Imagine," McConnell went on, warming up to his theme, "that they could destroy our critical infrastructure."

Obama, seeing where the director was headed, said, as if completing his sentence, "That would be problematic for the nation."

"*That's* the danger," McConnell said, and then he stepped into his well-rehearsed summation of the nation's vulnerabilities and the ability of many powers, not just China, to exploit them.

At its conclusion, Obama told McConnell to come see him again in the first week of his presidency.

In fact, McConnell next saw Obama in his transition office, on December 8, midway between his election-night victory and inauguration day. He brought with him his aide, Melissa Hathaway, who briefly outlined the Comprehensive National Cybersecurity

Initiative that she'd written for Bush but that hadn't yet been implemented. Obama told her to start thinking about a sixty-day review of U.S. cyber policy.

The review hit a slight delay. Cyber was hardly the most urgent issue on the new president's agenda. He first ordered another campaign aide, a former CIA analyst named Bruce Riedel, to write a sixty-day review of U.S. policy in Afghanistan. Then there was the matter of solving the banking crash, the collapse of the auto industry, and the worst economic crisis since the Great Depression.

Still, on February 9, just three weeks into his term, not too far behind schedule, Obama publicly announced the sixty-day cyber review and presented Hathaway as its chair. It took longer than sixty days to complete—it took 109 days—but on May 29, she and her interagency group issued their seventy-two-page document, titled *Cyberspace Policy Review: Assuring a Trusted and Resilient Information and Communications Infrastructure*.

It read uncannily like the reports, reviews, and directives that had come before it, and even referred to several of them by name, among them Bush's NSPD-54 and his *National Strategy to Secure Cyberspace*, the Marsh Report, a few Defense Science Board studies, and Senator Nunn's hearings. There was little new to say about the subject; but few of the old things had ever been officially adopted, so no one had heard of them—outside the coterie of experts who'd been following the cycles for years or decades—and it was, therefore, no redundancy for Hathaway to re-recite the same old problems and remedies.

Once again, then, there was the preface noting the ubiquity of cyberspace, its "strategic vulnerabilities," the "great risks" posed to "critical infrastructure" and "sensitive military information." There was the bit about the "overlapping authorities" among federal agencies, the need for a "national dialogue," an "action plan" for "information sharing" with "public-private partnerships," and, finally, the proposed appointment of a "cybersecurity policy official" in the

White House—a position that Hathaway assumed she would hold, just as Dick Clarke designated himself the "national coordinator" in a similar document for Bill Clinton.

But from the outset, Hathaway ran into obstacles. White House staffers disdained her as "prickly" and "sharp-elbowed," diatribes commonly hurled at women—Hathaway was blond, attractive, and barely forty—for behavior that would be tolerated as merely aggressive, or even normal, possibly admirable, in men. Hathaway's elbows were certainly less sharp than Clarke's, but Clarke was a master of office politics, cultivating protectors at the highest echelons and allies throughout the bureaucracy. Hathaway had only one protector, Mike McConnell, and when Obama replaced him in the first week of his presidency, she was left with no cover.

There was another problem, one that Clarke had also faced. Hathaway's review noted that private companies owned most of the pathways of cyberspace and thus must "share the responsibility" for its security—a line that triggered reflexive fears of government regulation, still the nastiest word in the book among the executives of Silicon Valley. Obama's brash economic adviser, Lawrence Summers, took industry's side in this dispute, insisting that, especially during what had come to be called the Great Recession, the engines of economic growth must not be constrained. (As Clinton's treasury secretary, Summers had been Clarke's bête noire when he tried to push for regulations, as well.)

Between the prominence of economic concerns and her own bureaucratic isolation, Hathaway and her portfolio took a tumble. She was gone by August and sidelined well before then.

Yet Obama didn't ignore Hathaway's concerns. On May 29, the same day that she released her review, he spoke for seventeen minutes in the East Room of the White House on cyberspace, its central place in modern life, and "this cyber threat" as "one of the most serious economic and national security challenges we face as a nation."

He spoke not just from a script but also from personal experience. Born in 1961, near the end of the baby boom (unlike Bush and Clinton, who were born fifteen years earlier, at the boom's onset), Obama was the first American president who surfed through cyberspace in his daily life. (When the Secret Service demanded that he give up his BlackBerry for security reasons, Obama resisted; as a compromise, the NSA Information Assurance Directorate built him a one-of-a-kind BlackBerry, equipped with state-of-the-art encryption, shielding, and a few other highly classified tricks.) And he was the first president whose campaign records had been hacked by a foreign power. Obama understood the stakes.

But something else stirred his concerns. A few days before inauguration day, President Bush had briefed him on two covert operations that he hoped Obama would continue. One concerned secret drone strikes against al Qaeda militants in Pakistan. The other involved a very tightly held, astonishingly bold cyber offensive campaign—code-named Operation Olympic Games, later known as Stuxnet—to delay and disable what seemed to be a nuclear weapons program in Iran.

Coming so soon after Mike McConnell's briefing on America's vulnerability to cyber attacks, this disclosure switched on a different light bulb from the one that had flashed in the heads of presidents, senior officials, and advisers who'd been exposed to the subject in the decades before. It was the obverse of the usual lesson: what the enemy might someday do to us, we can now do to the enemy.

CHAPTER 12

"SOMEBODY HAS CROSSED THE RUBICON"

G EORGE W. BUSH personally briefed Barack Obama on Olympic Games, rather than leave the task to an intelligence official, because, like all cyber operations, it required presidential authorization. After his swearing-in, Obama would have to renew the program explicitly or let it die; so Bush made a forceful plea to let it roll forward. The program, he told his successor, could mean the difference between a war with Iran and a chance for peace.

The operation had been set in motion a few years earlier, in 2006, midway through Bush's second term as president, when Iranian scientists were detected installing centrifuges—the long, silvery paddles that churn uranium gas at supersonic speeds—at a reactor in Natanz. The avowed purpose was to generate electrical power, but if the centrifuges cascaded in large enough quantities for a long enough time, the same process could make the stuff of nuclear weapons.

Vice President Cheney advocated launching air strikes on the Natanz reactor, as did certain Israelis, who viewed the prospect of a nuclear-armed Iran as an existential threat. Bush might have gone for the idea a few years earlier, but he was tiring of Cheney's relentless hawkishness. Bob Gates, the new defense secretary, had persuaded Bush that going to war against a *third* Muslim country, while the two in Afghanistan and Iraq were still raging, would be bad for national security. And so Bush was looking for a "third option"—something in between air strikes and doing nothing.

The answer came from Fort Meade—or, more precisely, from the decades-long history of studies, simulations, war games, and clandestine real-life excursions in counter-C2 warfare, information warfare, and cyber warfare, whose innovations and operators were now all centered at Fort Meade.

Like most reactors, Natanz operated with remote computer controls, and it was by now widely known—in a few months, it would be demonstrated with the Aurora Generator Test at the Idaho National Laboratory—that these controls could be hacked and manipulated in a cyber attack.

With this in mind, Keith Alexander, the NSA director, proposed launching a cyber attack on the controls of the Natanz reactor.

Already, his SIGINT teams had discovered vulnerabilities in the computers controlling the reactor and had prowled through their network, scoping out its dimensions, functions, and features, and finding still more vulnerabilities. This was digital age espionage, CNE—Computer Network Exploitation—so it didn't require the president's approval. For the next step, CNA, Computer Network Attack, the commander-in-chief's formal go-ahead would be needed. In preparation for the green light, Alexander laid out the rudiments of a plan.

In their probes, the NSA SIGINT teams had discovered that the software controlling the Natanz centrifuges was designed by Siemens, a large German company that manufactured PLCs—programmable logic controllers—for industrial systems worldwide.

The challenge was to devise a worm that would infect the Natanz system but no other Siemens systems elsewhere, in case the worm spread, as worms sometimes did.

Bush was desperate for some way out; this might be it; there was no harm in trying. So he told Alexander to proceed.

This would be a huge operation, a joint effort by the NSA, CIA, and Israel's cyber war bureau, Unit 8200. Meanwhile, Alexander got the operation going with a simpler trick. The Iranians had installed devices called uninterruptible power supplies on the generators that pumped electricity into Natanz, to prevent the sorts of spikes or dips in voltage that could damage the spinning centrifuges. It was easy to hack into these supplies. One day, the voltage spiked, and fifty centrifuges exploded. The power supplies had been ordered from Turkey; the Iranians suspected sabotage and turned to another supplier, thinking that would fix the problem. They were right about the sabotage, but not about its source.

Shutting down the reactor by messing with its power supplies was a one-time move. While the Iranians made the fix, the NSA prepared the more durable, devastating potion.

Most of this work was done by the elite hackers in TAO, the Office of Tailored Access Operations, whose technical skills and resources had swelled in the decade since Ken Minihan set aside a corner of the SIGINT Directorate to let a new cadre of computer geeks find their footing. For Olympic Games, they took some of their boldest inventions—which astounded even the most jaded SIGINT veterans who were let in on the secret—and combined them into a single super-worm called Flame.

A multipurpose piece of malware that took up 650,000 lines of code (nearly 4,000 times larger than a typical hacker tool), Flame—once it infected a computer—could swipe files, monitor keystrokes and screens, turn on the machine's microphone to record conversations nearby, turn on its Bluetooth function to steal data from most

smart phones within twenty meters, among other tricks, all from
NSA command centers across the globe.

To get inside the controls at Natanz, TAO hackers developed
malware to exploit five separate vulnerabilities that no one had previously discovered—five zero-day exploits—in the Windows operating system of the Siemens controllers. Exploiting one of these
vulnerabilities, in the keyboard file, gave TAO special user privileges
throughout a computer's functions. Another allowed access to all the
computers that shared an infected printer.

The idea was to hack into the Siemens machines controlling the
valves that pumped uranium gas into the centrifuges. Once this was
accomplished, TAO would manipulate the valves, turning them way
up, overloading the centrifuges, causing them to burst.

It took eight months for the NSA to devise this plan and design
the worm to carry it out. Now the worm had to be tested. Keith
Alexander and Robert Gates cooked up an experiment, in which
the technical side of the intelligence community would construct
a cascade of centrifuges, identical to those used at Natanz, and set
them up in a large chamber at one of the Department of Energy's
weapons labs. The exercise was similar to the Aurora test, which
took place around the same time, proving that an electrical generator
could be destroyed through strictly cyber means. The Natanz simulation yielded similar results: the centrifuges were sent spinning at
five times their normal speed, until they broke apart.

At the next meeting on the subject in the White House Situation
Room, the rubble from one of those centrifuges was placed on the
table in front of President Bush. He gave the go-ahead to try it out
on the real thing.

There was one more challenge: after the Iranians replaced the
sabotaged power supplies from Turkey, they took the additional pre-
caution of taking the reactor's computers offline. They knew about
the vulnerability of digital controls, and they'd read that surround-

ing computers with an air gap—cutting them off from the Internet, making their operations autonomous—was one way to eliminate the risks: if the system worked on a closed network, if hackers couldn't get into it, they couldn't corrupt, degrade, or destroy it, either.

What the Iranians didn't know was that the hackers of TAO had long ago figured out how to leap across air gaps. First, they'd penetrated a network near the air-gapped target; while navigating its pathways, they would usually find some link or portal that the security programmers had overlooked. If that path led nowhere, they would turn to their partners in the CIA's Information Operations Center. A decade earlier, during the campaign against Serbian President Slobodan Milosevic, IOC spies gained entry to Belgrade's telephone exchange and planted devices, which the NSA's SIGINT teams then hacked, giving them full access to the nation's phone system. These sorts of joint operations had blossomed with the growth of TAO.

The NSA also enjoyed close relations with Israel's Unit 8200, which was tight with the human spies of Mossad. If it needed access to a machine or a self-contained network that wasn't hooked up to the Internet, it could call on any of several collaborators—IOC, Unit 8200, the local spy services, or certain defense contractors in a number of allied nations—to plant a transmitter or beacon that TAO could home in on.

In Olympic Games, someone would install the malware by physically inserting a thumb drive into a computer (or a printer that several computers were using) on the premises—in much the same way that, around this same time, Russian cyber warriors hacked into U.S. Central Command's classified networks in Afghanistan, the intrusion that the NSA detected and repelled in Operation Buckshot Yankee.

Not only would the malware take over the Natanz reactor's valve pumps, it would also conceal the intrusion from the reactor's overseers. Ordinarily, the valve controls would send out an alert when the flow of uranium rapidly accelerated. But the malware allowed TAO to intercept the alert and to replace it with a false signal, indicating that everything was fine.

The worm could have been designed to destroy every centrifuge, but that would arouse suspicions of sabotage. A better course, its architects figured, would be to damage just enough centrifuges to make the Iranians blame the failures on human error or poor design. They would then fire perfectly good scientists and replace perfectly good equipment, setting back their nuclear program still further.

In this sense, Operation Olympic Games was a classic campaign of information warfare: the target wasn't just the Iranians' nuclear program but also the Iranians' confidence—in their sensors, their equipment, and themselves.

The plan was ready to go, but George Bush's time in office was running out. It was up to Barack Obama.

To Bush, the plan, just like the one to send fake email to Iraqi insurgents, was a no-brainer. It made sense to Obama, too. From the outset of his presidency, Obama articulated, and usually followed, a philosophy on the use of force: he was willing to take military action, if national interests demanded it and if the risks were fairly low; but unless *vital* interests were at stake, he was averse to sending in thousands of American troops, especially given the waste and drain of the two wars he inherited in Afghanistan and Iraq. The two secret programs that Bush pressed him to continue—drone strikes against jihadists and cyber sabotage of a uranium-enrichment plant in Iran—fit Obama's comfort zone: both served a national interest, and neither risked American lives.

Once in the White House, Obama expressed a few qualms about the plan: he wanted assurances that, when the worm infected the Natanz reactor, it wouldn't also put out the lights in nearby power plants, hospitals, or other civilian facilities.

His briefers conceded that worms could spread, but this particular worm was programmed to look for the specific Siemens software; if it drifted far afield, and the unintended targets didn't have the software, it wouldn't inflict any damage.

Gates, who'd been kept on by Obama and was already a major influence on his thinking, encouraged the new president to renew the go-ahead. Obama saw no reason not to.

Not quite one month after he took office, the worm had its first success: a cascade of centrifuges at Natanz sped out of control, and several of them shattered. Obama phoned Bush to tell him the covert program they'd discussed was working out.

In March, the NSA shifted its approach. In the first phase, the operation hacked into the valves controlling the rate at which uranium gas flowed into the centrifuges. In the second phase, the attack went after the devices—known as frequency converters—that controlled how quickly the centrifuges rotated. The normal speed ranged from about 800 to 1,200 cycles per second; the worm gradually sped them up to 1,410 cycles, at which point several of the centrifuges flew apart. Or, sometimes, it slowed down the converters, over a period of several weeks, to as few as 2 cycles per second: as a result, the uranium gas couldn't exit the centrifuge quickly enough; the imbalance would cause vibrations, which severely damaged the centrifuge in a different way.

Regardless of the technique, the worm also fed false data to the system's monitors, so that, to the Iranian scientists watching them, everything seemed normal—and, when disaster struck, they couldn't figure out what had happened. They'd experienced technical problems with centrifuges from the program's outset; this seemed—and the NSA designed the worm to *make* it seem—like more of the same, but with more intense and frequent disruptions.

By the start of 2010, nearly a quarter of Iran's centrifuges—about 2,000 out of 8,700—were damaged beyond repair. U.S. intelligence analysts estimated a setback in Iran's enrichment program of two to three years.

Then, early that summer, it all went wrong. President Obama—who'd been briefed on every detail and alerted to every success or breakdown—was told by his advisers that the worm was out of the box: for reasons not entirely clear, it had jumped from one computer to another, way outside the Natanz network, then to another network outside that. It wouldn't wreak damage—as the briefers had told him before, it was programmed to shut down if it didn't see a particular Siemens controller—but it would get noticed: the Iranians would eventually find out what had been going on; Olympic Games was on the verge of being blown.

Almost at once, some of the world's top software security firms—Symantec in California, VirusBlokAda in Belarus, Kaspersky Lab in Russia—started detecting a strange virus randomly popping up around the world. At first, they didn't know its origins or its purpose; but probing its roots, parsing its code, and gauging its size, they realized they'd hit upon one of the most elaborate, sophisticated worms of all time. Microsoft issued an advisory to its customers, and, forming an anagram from the first few letters on the code, called the virus "Stuxnet"—a name that caught on.

By August, Symantec had uncovered enough evidence to release a statement of its own, warning that Stuxnet was designed not for mischievous hacking or even for espionage, but rather for sabotage. In September, a German security researcher named Ralph Langner inferred, from the available facts, that someone was trying to disable the Natanz nuclear reactor in Iran and that Israelis were probably involved.

At that point, some of the American software sleuths were horrified: Had they just helped expose a highly classified U.S. intelligence operation? They couldn't have known at the time, but their curiosity—and their professional obligation to inform the public about a loose and possibly damaging computer virus—did have that effect. Shortly after Symantec's statement, even before Langner's educated guess about Stuxnet's true aim, the Iranians drew the proper inference (so *this* was why their centrifuges were spinning out of control) and cut off all links between the Natanz plant and the Siemens controllers.

When Obama learned of the exposure at a meeting in the White House, he asked his top advisers whether they should shut down the operation. Told that it was still causing damage, despite Iranian countermeasures, he ordered the NSA to intensify the program— sending the centrifuges into wilder contortions, speeding them up, then slowing them down—with no concerns about detection, since its cover was already blown.

The postmortem indicated that, in the weeks *after* the exposure, another 1,000 centrifuges, out of the remaining 5,000, were taken out of commission.

———

Even after Olympic Games came to an end, the art and science of CNA—Computer Network Attack—pushed on ahead. In fact, by the end of October, when U.S. Cyber Command achieved full readiness for operations, CNA emerged as a consuming, even dominant, activity at Fort Meade.

A year earlier, anticipating Robert Gates's directive creating Cyber Command, the chairman of the Joint Chiefs of Staff, General Peter Pace, issued a classified document, *National Military Strategy for Cyber Operations*, which expressed the need for "offensive capabilities in cyber space to gain and maintain the initiative."

General Alexander, now CyberCom commander as well as the NSA director, was setting up forty "cyber-offensive teams"—twenty-seven for the U.S. combatant commands (Central Command, Pacific Command, European Command, and so forth) and thirteen engaged in the defense of networks, mainly Defense Department networks, at home. Part of this latter mission involved monitoring the networks; thanks to the work of the previous decade, starting with the Air Force Information Warfare Center, then gradually extending to the other services, the military networks had so few access points to the Internet—just twenty by this time, cut to eight in the next few years—that Alexander's teams *could* detect and repel attacks across the transom. But defending networks also meant going on the offensive, through the deliberately ambiguous concept of CNE, Computer Network Exploitation, which could be both a form of "active defense" and preparation for CNA—Computer Network *Attack*.

Some officials deep inside the national security establishment were concerned about this trend. The military—the nation—was rapidly adopting a new form of warfare, had assembled and *used* a new kind of weapon; but this was all being done in great secrecy, inside the nation's most secretive intelligence agency, and it was clear, even to those with a glimpse of its inner workings, that no one had thought through the implications of this new kind of weapon and new vision of war.

During the planning for Stuxnet, there had been debates, within the Bush and Obama administrations, over the precedent that the attack might establish. For more than a decade, dozens of panels and commissions had warned that America's critical infrastructure was vulnerable to a cyber attack—and now *America* was launching the first cyber attack on *another* nation's critical infrastructure. Almost no one outright opposed the Stuxnet program: if it could keep Iran from developing nuclear weapons, it was worth the risk; but several officials realized that it *was* a risk, that the dangers of blowback were inescapable and immense.

The United States wasn't alone on this cyber rocket ship, after all. Ever since their penetration of Defense Department sites a decade earlier, in Operation Moonlight Maze, the Russians had been ramping up their capabilities to exploit and attack computer networks. The Chinese had joined the club in 2001 and soon grew adept at penetrating sensitive (though, as far as anyone knew, unclassified) networks of dozens of American military commands, facilities, and laboratories. In Obama's first year as president, around the Fourth of July, the North Koreans—whose citizens barely had electricity—launched a massive denial-of-service attack, shutting down websites of the Departments of Homeland Security, Treasury, Transportation, the Secret Service, the Federal Trade Commission, the New York Stock Exchange, and NASDAQ, as well as dozens of South Korean banks, affecting at least 60,000, possibly as many as 160,000 computers.

Stuxnet spurred the Iranians to create their own cyber war unit, which took off at still greater levels of funding a year and a half later, in the spring of 2012, when, in a follow-up attack, the NSA's Flame virus—the massive, multipurpose malware from which Olympic Games had derived—wiped out nearly every hard drive at Iran's oil ministry and at the Iranian National Oil Company. Four months after that, Iran fired back with its own Shamoon virus, wiping out 30,000 hard drives (basically, every hard drive in every workstation) at Saudi Aramco, the joint U.S.-Saudi Arabian oil company, and planting, on every one of its computer monitors, the image of a burning American flag.

Keith Alexander learned, from communications intercepts, that the Iranians had expressly developed and launched Shamoon as retaliation for Stuxnet and Flame. On his way to a conference with GCHQ, the NSA's British counterpart, he read a talking points memo, written by an aide, noting that, with Shamoon and several other recent cyber attacks on Western banks, the Iranians had "demonstrated a clear ability to learn from the capabilities and actions of others"—namely, those of the NSA and of Israel's Unit 8200.

It was the latest, most dramatic illustration of what agency analysts and directors had been predicting for decades: what we can
do to them, they can someday do to us—except that "someday"
was *now*.

Alexander's term as NSA director was coinciding with—and Alexander himself had been fostering—not only the advancement of
cyber weapons and the onset of *physically* destructive cyber attacks,
but also the early spirals of a cyber arms *race*. What to do about it?
This, too, was a question that no one had thought through, at even
the most basic level.

When Bob Gates became secretary of defense, back at the end
of 2006, he was so stunned by the volume of attempted intrusions
into American military networks—his briefings listed dozens, sometimes hundreds, every *day*—that he wrote a memo to the Pentagon's
deputy general counsel. At what point, he asked, did a cyber attack
constitute an act of war under international law?

He didn't receive a reply until the last day of 2008, almost two
years later. The counsel wrote that, yes, a cyber attack *might* rise to the
level that called for a military response—it *could* be deemed an act of
armed aggression, under certain circumstances—but what those circumstances were, where the line should be drawn, even the criteria
for drawing that line, were matters for policymakers, not lawyers, to
address. Gates took the reply as an evasion, not an answer.

One obstacle to a clearer answer—to clearer thinking, generally—
was that everything about cyber war lay encrusted in secrecy: its roots
were planted, and its fruits were ripening, in an agency whose very
existence had once been highly classified and whose operations were
still as tightly held as any in government.

This culture of total secrecy had a certain logic back when
SIGINT was strictly an intelligence tool: the big secret was that the
NSA had broken some adversary's code; if that was revealed, the
adversary would simply change the code; the agency would have to

start all over, and until it broke the new code, national security could be damaged; in wartime, a battle might be lost.

But now that the NSA director was also a four-star commander, and now that SIGINT had been harnessed into a weapon of destruction, something like a remote-control bomb, questions were raised and debates were warranted, for reasons having to do not only with morality but with the new weapon's strategic usefulness—its precise effects, side effects, and consequences.

General Michael Hayden, the former NSA director, had moved over to Langley, as director of the CIA, when President Bush gave the go-ahead on Olympic Games. (He was removed from that post when Obama came to the White House, so he had no role in the actual operation.) Two years after Stuxnet came crashing to a halt, when details about it were leaked to the mainstream press, Hayden— by now retired from the military—voiced in public the same concerns that he and others had debated in the White House Situation Room.

"Previous cyber-attacks had effects limited to other computers," Hayden told a reporter. "This is the first attack of a major nature in which a cyber-attack was used to effect physical destruction. And no matter what you think of the effects—and I think destroying a cascade of Iranian centrifuges is an unalloyed good—you can't help but describe it as an attack on critical infrastructure."

He went on: "Somebody has crossed the Rubicon. We've got a legion on the other side of the river now." Something had shifted in the nature and calculation of warfare, just as it had after the United States dropped atom bombs on Hiroshima and Nagasaki at the end of World War II. "I don't want to pretend it's the same effect," Hayden said, "but in one sense at least, it's August 1945."

For the first two decades after Hiroshima, the United States enjoyed vast numerical superiority—for some of that time, a monopoly—in nuclear weapons. But on the cusp of a new era in cyber war, it was a known fact that many other nations had cyber war units, and America was far more vulnerable in this kind of war than any likely adversary, than any other country on the planet, because it relied far more heavily on vulnerable computer networks—in its weapons systems, its financial systems, its vital critical infrastructures.

If America, or U.S. Cyber Command, wanted to wage cyber war, it would do so from inside a glass house.

There was another difference between the two kinds of new weapons, besides the scale of damage they could inflict: nuclear weapons were out there, in public; certain aspects of their production or the exact size of their stockpile were classified, but everyone knew who had them, everyone had seen the photos and the film clips, showing what they could do, if they were used; and if they were used, everyone would know who had launched them.

Cyber weapons—their existence, their use, and the policies surrounding them—were still secret. It *seemed* that the United States and Israel sabotaged the Natanz reactor, that Iran wiped out Saudi Aramco's hard drives, and that North Korea unleashed the denial-of-service attacks on U.S. websites and South Korean banks. But no one took credit for the assaults; and while the forensic analysts who traced the attacks were confident in their assessments, they didn't—they couldn't—boast the same slam-dunk certainty as a physicist tracking the arc of a ballistic missile's trajectory.

This extreme secrecy extended not only to the mass public but also inside the government, even among most officials with high-level security clearances. Back in May 2007, shortly after he briefed George W. Bush on the plan to launch cyber attacks against Iraqi insurgents, Mike McConnell, then the director of national intelligence, hammered out an accord with senior officials in the Pentagon, the NSA, the CIA, and the attorney general's office, titled "Trilateral Memorandum of Agreement Among the Department of

Defense, the Department of Justice, and the Intelligence Community Regarding Computer Network Attack and Computer Network Exploitation Activities." But, apart from the requirement that cyber offensive operations needed presidential approval, there were no formal procedures or protocols for top policy advisers and policymakers to assess the aims, risks, benefits, or consequences of such attacks.

To fill that vast blank, President Obama ordered the drafting of a new presidential policy directive, PPD-20, titled "U.S. Cyber Operations Policy," which he signed in October 2012, a few months after the first big press leaks about Stuxnet.

Eighteen pages long, it was the most explicit, detailed directive of its kind. In one sense, its approach was more cautious than its predecessors. It noted, for instance, in an implied (but unstated) reference to Stuxnet's unraveling, that the effects of a cyber attack can spread to "locations other than the intended target, with potential unintended or collateral consequences that may affect U.S. national interests." And it established an interagency Cyber Operations Policy Working Group to ensure that such side effects, along with other broad policy issues, were weighed before an attack was launched.

But the main intent and impact of PPD-20 was to institutionalize cyber attacks as an integral tool of American diplomacy and war. It stated that the relevant departments and agencies "shall identify potential targets of national importance" against which cyber attacks "can offer a favorable balance of effectiveness and risk as compared to other instruments of national power." Specifically, the secretary of defense, director of national intelligence, and director of the CIA— in coordination with the attorney general, secretary of state, secretary of homeland security, and relevant heads of the intelligence community—"shall prepare, for approval by the President . . . a plan that identifies potential systems, processes, and infrastructure against which the United States should establish and maintain [cyber offensive] capabilities; proposes circumstances under which [they] might be used; and proposes necessary resourcing and steps that would be needed for implementation, review, and updates as U.S. national security needs change."

Cyber options were to be systematically analyzed, preplanned, and woven into broader war plans, in much the same way that nuclear options had been during the Cold War.

Also, as with nuclear options, the directive required "specific Presidential approval" for any cyber operation deemed "reasonably likely to result in 'significant consequences' "—those last two words defined to include "loss of life, significant responsive actions against the United States, significant damage to property, serious adverse U.S. foreign policy consequences, or serious economic impact to the United States"—though an exception was made, allowing a relevant agency or department head to launch an attack without presidential approval in case of an emergency.

However, unlike nuclear options, the plans for cyber operations were not intended to lie dormant until the *ultimate* conflict; they were meant to be executed, and fairly frequently. The agency and department heads conducting these attacks, the directive said, "shall report annually on the use and effectiveness of operations of the previous year to the President, through the National Security Adviser."

No time was wasted in getting these plans up and ready. An action report on the directive noted that the secretary of defense, director of national intelligence, and CIA director briefed an NSC Deputies meeting on the scope of their plans in April 2013, six months after PPD-20 was signed.

PPD-20 was classified TOP SECRET/NOFORN, meaning it could not be shared with foreign officials; the document's very existence was highly classified. But it was addressed to the heads of all the relevant agencies and departments, and to the vice president and top White House aides. In other words, the subject was getting discussed, not only in these elite circles, but also—with Stuxnet out in the open—among the public. Gingerly, officials began to acknowledge, in broad general terms, the existence and concept of cyber offensive operations.

General James Cartwright, who'd recently retired as vice chairman of the Joint Chiefs of Staff and who, before then, had been head of U.S. Strategic Command, which had nominal control over cyber operations, told a reporter covering Stuxnet that the extreme secrecy surrounding the topic had hurt American interests. "You can't have something that's a secret be a deterrent," he said, "because if you don't know it's there, it doesn't scare you."

Some officers dismissed Cartwright's logic: the Russians and Chinese knew what we had, just as much as we knew what they had. Still, others agreed that it might be time to open up a little bit.

In October, the same month that PPD-20 was signed, the NSA declassified a fifteen-year-old issue of *Cryptolog*, the agency's in-house journal, dealing with the history of information warfare. The special issue had been published in the spring of 1997, its contents stamped TOP SECRET UMBRA, denoting the most sensitive level of material dealing with communications intelligence. One of the articles, written by William Black, the agency's top official for information warfare at the time, noted that the secretary of defense had delegated to the NSA "the authority to develop Computer Network Attack (CNA) techniques." In a footnote, Black cited a Defense Department directive from the year before, defining CNA as "operations to disrupt, deny, degrade, or destroy information resident in computers and computer networks, or the computers and networks themselves."

This was remarkably similar to the way Obama's PPD-20 defined "cyber effect"—as the "manipulation, disruption, denial, degradation, or destruction of computers, information or communications systems, networks, physical or virtual infrastructure controlled by computers or information systems, or information resident therein."

In this sense, PPD-20 was expressing, in somewhat more detailed language, an idea that had been around since William Perry's counter command-control warfare in the late 1970s.

After all those decades, the declassified *Cryptolog* article marked the first time that the term CNA, or such a precise definition of the concept, had appeared in a public document.

Within the Air Force, which had always been the military service most active in cyberspace, senior officers started writing a policy statement acknowledging its CNA capabilities, with the intent of releasing the paper to the public.

But then, just as they were finishing a draft, the hammer came down. Leon Panetta, a former Democratic congressman and budget director who'd replaced a fatigued Robert Gates as Obama's secretary of defense, issued a memo forbidding any further references to America's CNA programs.

Obama had decided to confront the Chinese directly on their rampant penetrations of U.S. computer networks. And Panetta didn't want his officers to supply the evidence that might help the Chinese accuse the American president of hypocrisy.

CHAPTER 13

SHADY RATs

O N March 11, 2013, Thomas Donilon, President Obama's national security adviser, gave a speech at the Asia Society on Manhattan's Upper East Side. Much of it was boilerplate: a recitation of the administration's policy of "rebalancing its global posture" away from the ancient battles of the Middle East and toward the "dynamic" region of Asia-Pacific as a force for growth and prosperity.

But about two thirds of the way through the speech, Donilon broke new diplomatic ground. After listing a couple of "challenges" facing U.S.-China relations, he said, "Another such issue is cyber security," adding that Chinese aggression in this realm had "moved to the forefront of our agenda."

American corporations, he went on, were increasingly concerned "about sophisticated, targeted theft of confidential business information and proprietary technologies through cyber intrusions emanating from China on an unprecedented scale."

Then Donilon raised the stakes higher. "From the president on down," he said, "this has become a key point of concern and discussion with China at all levels of our governments. And it will continue to be. The United States will do all it must to protect our national networks, critical infrastructure, and our valuable public and private sector property."

The Obama administration, he said, wanted Beijing to do two things: first, to recognize "the urgency and scope of this problem and the risk it poses—to international trade, to the reputation of Chinese industry, and to our overall relations"; second, to "take serious steps to investigate and put a stop to these activities."

The first demand was a borderline threat: change your ways or risk a rupture of our relations. The second was an attempt to give Chinese leaders a face-saving way out, an opportunity for them to blame the hacking on hooligans and "take serious steps" to halt it.

In fact, Donilon and every other official with a high-level security clearance *knew* that the culprit, in these intrusions, was no gang of freelance hackers but rather the Chinese government itself—specifically, the Second Bureau of the Third Department of the People's Liberation Army's General Staff, also known as PLA Unit 61398, which was headquartered in a white, twelve-story office building on the outskirts of Shanghai.

Since the start of his presidency, Obama had raised the issue repeatedly but quietly—in part to protect intelligence sources and methods, in part because he wanted to improve relations with China and figured a confrontation over cyber theft would impede those efforts. His diplomats brought it up, as a side issue, at every one of their annual Asian-American "strategic and economic dialogue" sessions, beginning with Obama's first, in 2009. On none of those occasions did the Chinese delegates bite: to the extent they replied at all, they agreed that the international community must put a stop to this banditry; if an American diplomat brought up China's own involvement in cyber hacking, they waved off the accusation.

Then, on February 18, Mandiant, a leading computer-security firm, with headquarters in Alexandria, Virginia, published a sixty-page report identifying PLA Unit 61398 as one of the world's most prodigious cyber hackers. Over the previous seven years, the report stated, the Shanghai hackers had been responsible for at least 141 successful cyber intrusions in twenty major industrial sectors, including defense contractors, waterworks, oil and gas pipelines, and other critical infrastructures. On average, these hackers lingered inside a targeted network for a full year—in one case, for four years and ten months—before they were detected. During one particularly unimpeded operation, they filched 6.5 terabytes of data from a single company in a ten-month period.

Kevin Mandia, the founder and chief executive of Mandiant, had been one of the Air Force cyber crime investigators who, fifteen years earlier, had nailed Moscow as the culprit of Moonlight Maze, the first serious foreign hacking of Defense Department computers. Mandiant's chief security officer, Richard Bejtlich, had been, around the same time, a computer network defense specialist at the Air Force Information Warfare Center, which installed the first network security monitors to detect and track penetrations of military computers. The monitoring system that Mandia and Bejtlich built at Mandiant was based on the system that the Air Force used in San Antonio.

While putting together his report on Unit 61398, Mandia was contracted by *The New York Times* to investigate the hacking of its news division. As that probe progressed (it turned out that the hacker was a different Chinese government organization), he and the paper's publishers discussed a possible long-term business arrangement, so he gave them an advance copy of the report on the Shanghai unit. The *Times* ran a long front-page story summarizing its contents.

China's foreign affairs ministry denounced the allegation as "irresponsible," "unprofessional," and "not helpful for the resolution of the relevant problem," adding, in the brisk denial that its officials had always recited in meetings with American diplomats, "China resolutely opposes hacking actions."

In fact, however, the Chinese had been hacking, with growing profligacy, for more than a decade. A senior U.S. intelligence official had once muttered at an NSC meeting that at least the Russians tried to keep their cyber activity secret; the Chinese just did it everywhere, out in the open, as if they didn't care whether anyone noticed.

As early as 2001, in an operation that American intelligence agencies dubbed Titan Rain, China's cyber warriors hacked into the networks of several Western military commands, government agencies, defense corporations, and research labs, using techniques reminiscent of the Russians' Moonlight Maze operation.

Around the same time, the Third Department of the PLA's General Staff, which later created Unit 61398, adopted a new doctrine that it called "information confrontation." Departments of "information-security research" were set up in more than fifty Chinese universities. By the end of the decade, the Chinese army started to incorporate cyber tools and techniques in exercises with names like "Iron Fist" and "Mission Attack"; one scenario had the PLA hacking into U.S. Navy and Air Force command-control networks in an attempt to impede their response to an occupation of Taiwan.

In short, the Chinese were emulating the American doctrine of "information warfare"—illustrating, once more, the lesson learned by many who found the cyber arts at first alluring, then alarming: what we could do to an adversary, an adversary could do to us.

There was one big difference in the Chinese cyber attacks: they were engaging not just in espionage and battlefield preparation, but also in the theft of trade secrets, intellectual property, and cash.

In 2006, if not sooner, various cyber bureaus of the Chinese military started hacking into a vast range of enterprises worldwide. The campaign began with a series of raids on defense contractors, notably a massive hack of Lockheed Martin, where China stole tens of millions of documents on the company's F-35 Joint Strike Fighter aircraft. None of the files were classified, but they contained data and blueprints on cockpit design, maintenance procedures, stealth technology, and other matters that could help the Chinese counter the plane in battle or, meanwhile, build their own F-35 knockoff (which they eventually did).

Colonel Gregory Rattray, a group commander in the Air Force Information Warfare Center (which had recently changed its name to the Air Force Information Operations Center), was particularly disturbed: not only by the scale of China's cyber raids but also by the passivity of American corporations. Rattray was an old hand in the field: he had written his doctoral dissertation on information warfare at the Fletcher School of Law and Diplomacy, worked on Richard Clarke's staff in the early years of George W. Bush's presidency, then, after Clarke resigned, stayed on as the White House director of cyber security.

In April 2007, Rattray summoned several executives from the largest U.S. defense contractors and informed them that they were living in a new world. The intelligence estimates that pinned the cyber attacks on China were highly classified; so, for one of his briefing slides, Rattray coined a term to describe the hacker's actions: "APT"—for advanced persistent threat. Its meaning was literal: the hacker was using sophisticated techniques; he was looking for specific information; and he was staying inside the system as long as necessary—weeks, even months—to find it. (The term caught on; six years later, Kevin Mandia titled his report *APT1*.)

The typical Chinese hack started off with a spear-phishing email to the target-company's employees. If just one employee clicked the email's attachment (and all it took was one), the computer would download a webpage crammed with malware, including a "Remote Access Trojan," known in the trade as a RAT. The RAT opened a door, allowing the intruder to roam the network, acquire the privileges of a systems administrator, and extract all the data he wanted.

They did this with economic enterprises of all kinds: banks, oil and gas pipelines, waterworks, health-care data managers—sometimes to steal secrets, sometimes to steal money, sometimes for motives that couldn't be ascertained.

McAfee, the anti-virus firm that discovered and tracked the Chinese hacking operation, called it Operation Shady RAT. Over a five-year period ending in 2011, when McAfee briefed the White House and Congress on its findings, Shady RAT stole data from more than seventy entities—government agencies and private firms—in fourteen countries, including the United States, Canada, several nations in Europe, and more in Asia, including many targets in Taiwan but, tellingly, none in the People's Republic of China.

President Obama didn't need McAfee to tell him about China's cyber spree; his intelligence agencies were filing similar reports. But the fact that a commercial anti-virus firm had tracked so much of the hacking, and released such a detailed report, made it hard to keep the issue locked up in the closet of diplomatic summits. The companies that were hacked would also have preferred to stay mum—no point upsetting customers and stockholders—but the word soon spread, and they reacted by pressuring the White House to do something, largely because, after all these decades of analyses and warnings, many of them still didn't know what to do themselves.

This was the setting that forced Obama's hand. After another Asia security summit, where his diplomats once again raised the issue and the Chinese once again denied involvement, he told Tom Donilon to deliver a speech that brought the issue out in the open. The Mandiant report—which had been published three weeks earlier—upped the pressure and accelerated the timetable, but the dynamics were already in motion.

One passage in Donilon's speech worried some midlevel officials, especially in the Pentagon. Characterizing cyber offensive raids as a violation of universal principles, even as something close to a cause for war, Donilon declared, "The international community cannot afford to tolerate any such activity from any country."

The Pentagon officials scratched their heads: "*any* such activity from *any* country"? The fact was, and everyone knew it, the United States engaged in this activity, too. Its targets were different: American intelligence agencies weren't stealing foreign companies' trade secrets or blueprints, much less their cash, mainly because they didn't need to be; such secrets or blueprints wouldn't have given American companies an advantage—they already *had* the advantage.

In NSC meetings on the topic, White House aides argued that this distinction was important: espionage for national security was an ancient, acceptable practice; but if the Chinese wanted to join the international economy, they had to respect the rights of property, including intellectual property. But other officials at these meetings wondered if there really was a difference. The NSA was hacking into Chinese networks to help defeat them in a war; China was hacking into American networks mainly to help enrich its economy. What made one form of hacking permissible and the other form intolerable?

Even if the White House aides had a point (and the Pentagon officials granted that they did), wasn't the administration skirting danger by going public with this criticism? Wouldn't it be too easy for the Chinese to release their own records, revealing that we were hacking them, too, and thus accuse us of hypocrisy? Part of what we were doing was defensive: penetrating their networks in order to follow them penetrating our networks; and we were penetrating these networks so deeply that, whenever the Chinese tried to hack into Defense Department systems (or, lately, those of several weapons contractors, too), the NSA was monitoring every step they took—it was monitoring what the Chinese were seeing on their own monitors. On a few occasions, the manufacturing secrets that the Chinese stole weren't real secrets at all; they were phony blueprints that the NSA had planted on certain sites as honey pots. But, to some extent, these cyber operations were *offensive* in nature: the United States was penetrating Chinese networks to prepare for battle, to exploit weaknesses and exert leverage, just as the Chinese were doing—just as every major power had always done in various realms of warfare.

The whole business of calling out China for hacking was particularly awkward, given the recent revelations about Stuxnet, to say nothing of Obama's recent (though still highly classified) signing of PPD-20, the presidential directive on cyber operations. Some of Obama's White House aides acknowledged a certain irony in the situation; it was one reason the administration refused to acknowledge having played a role in Stuxnet, long after the operation had been blown.

In May, Donilon flew to Beijing to make arrangements for a summit between President Obama and his Chinese counterpart, Xi Jinping. Donilon made it clear that cyber would be on the agenda and that, if necessary, Obama would let Xi in on just how much U.S. intelligence knew about Chinese practices. The summit was scheduled to take place in Rancho Mirage, California, at the estate of the late media tycoon Walter Annenberg, on Friday and Saturday, June 7 and 8, 2013.

On June 6, *The Washington Post* and *The Guardian* of London reported, in huge front-page stories, that, in a highly classified program known as PRISM, the NSA and Britain's GCHQ had long been mining data from nine Internet companies, usually under secret court orders and that, through this and other programs, the NSA was collecting telephone records of millions of American citizens. These were the first of many stories, published over the next several months by *The Guardian*, the *Post*, *Der Spiegel*, and eventually others, based on a massive trove of beyond-top-secret documents that NSA systems administrator Edward Snowden had swiped off his computer at the agency's facility in Oahu, Hawaii, and leaked to three journalists before fleeing to Hong Kong, where he met with two of them, Laura Poitras and Glenn Greenwald. (The other reporter, Barton Gellman, couldn't make the trip.)

The timing of the leak, coming on the eve of the Obama-Xi summit, was almost certainly happenstance—Snowden had been in touch with the reporters for months—but the effect was devastating. Obama brought up Chinese cyber theft; Xi took out a copy of *The Guardian*. From that point on, the Chinese retort to all American accusations on the subject shifted from "We don't do hacking" to "You do it a lot more than we do."

One week after the failed summit, as if to bolster Xi's position, Snowden—who, by this time, had revealed himself as the source in a dramatic video taped by Poitras in his hotel room—said, in an interview with Hong Kong's top newspaper, the *South China Morning Post*, that the NSA had launched more than 61,000 cyber operations, including attacks on hundreds of computers in Hong Kong and mainland China.

The *Morning Post* interview set off suspicions about Snowden's motives: he was no longer just blowing the whistle on NSA domestic surveillance; he was also blowing foreign intelligence operations. Soon came newspaper stories about NSA hacking into email traffic and mobile phone calls of Taliban insurgents on the eastern border of Afghanistan; an operation to gauge the loyalties of CIA recruits in Pakistan; email intercepts to assist intelligence assessments of events in Iran; and a surveillance program of cell phone calls "worldwide," intended to find and track associates of known terrorists.

One leak was the full, fifty-page catalogue of tools and techniques used by the elite hackers in the NSA's Office of Tailored Access Operations. No American or British newspaper published that document, though *Der Spiegel* did, in its print and online editions. Fort Meade's crown jewels were now scattered all over the global street, for interested parties everywhere to pick up. Even the material that no one published—and Snowden's cache amounted to tens of thousands of highly classified documents—could have been perused by any foreign intelligence agency with skilled cyber units. If the NSA and its Russian, Chinese, Iranian, French, and Israeli variants could hack into one another's computers, they could certainly hack into the computers of journalists, some of whom were less careful than others in guarding the cache. Once Snowden took his laptops out of the building in Oahu, its contents—encrypted or otherwise—were up for grabs.

But the leaks about foreign intelligence operations—the intercepts of email in Afghanistan and Pakistan, the TAO catalogue, and the like—were overshadowed, among American news readers, by the detailed accounts of domestic surveillance. It was these leaks that earned Snowden applause as a whistleblower and engulfed the NSA in a storm of controversy and protest unseen since the Church Committee hearings of the 1970s.

The Snowden papers unveiled a massive data-mining operation, more vast than any outsider had imagined. In effect, it was Keith Alexander's metadata experiment at Fort Belvoir writ very large—the realization of his philosophy about big data: collect and store everything, so that you can go back and search for patterns and clues of an imminent attack; when you're looking for a needle in a haystack, you need the whole haystack.

Under the surveillance system described in the Snowden documents, when the NSA found someone in contact with foreign terrorists, its analysts could go back and look at every phone number the suspect had called (and every number that had called the suspect) for the previous five years. The retrieval of all those associated numbers was called the first "hop." To widen the probe, analysts could then look at all the numbers that *those* people had called (the second hop) and, in a third hop, the numbers that *those* people had called.

The math suggested, at least potentially, a staggering level of surveillance. Imagine someone who had dialed the number of a known al Qaeda member, and assume that this person had phoned 100 other people over the previous five years. That would mean the NSA could start tracking not only the suspect's calls but also the calls of those 100 other people. If each of *those* people also called 100 people, the NSA—in the second hop—could track their calls, too, and that would put (100 times 100) 10,000 people on the agency's screen. In the third hop, the analysts could trace the calls of those 10,000 people and the calls that *they* had made—or (10,000 times 100) 1 million people.

In other words, the active surveillance of a single terrorist suspect could put a million people, possibly a million Americans, under the agency's watch. The revelation came as a shock, even to those who otherwise had few qualms about the occasional breach of personal privacy.

Following this disclosure, Keith Alexander gave several speeches and interviews, in which he emphasized that the NSA did not examine the *contents* of those calls or the names of the callers (that information was systematically excluded from the database) but rather only the *metadata*: the traffic patterns—which phone numbers called which other phone numbers—along with the dates, times, and durations of those calls.

But amid the dramatic news stories, an assurance from the director of the National Security Agency struck a weak chord: he might *say* that his agency didn't listen to these phone calls, but many wondered why they should believe him.

The distrust deepened when Obama's director of national intelligence, James Clapper, a retired Air Force lieutenant general and a veteran of various spy agencies, was caught in a lie. Back on March 12, three months before anyone had heard of metadata, PRISM, or Edward Snowden, Clapper testified at a public hearing of the Senate Select Committee on Intelligence. At one point, Senator Ron Wyden, Democrat from Oregon, asked him, "Does the NSA collect any type of data at all on millions or hundreds of millions of Americans?"

Clapper replied, "No, sir . . . not wittingly."

As a member of the select committee, Wyden had been read in on the NSA metadata program, so he knew that Clapper wasn't telling the truth. The day before, he'd given Clapper's office a heads-up on the question that he planned to ask. He knew that he'd be putting Clapper in a box: the correct answer to his question was "Yes," but Clapper would have a hard time saying so without making headlines; so Wyden wanted to give the director a chance to formulate an answer that addressed the issue without revealing too much. He was surprised that Clapper dealt with it by simply lying. After the hearing, Wyden had an aide approach Clapper to ask if he'd like to revise and extend his reply for the record; again to Wyden's surprise, Clapper declined. Wyden couldn't say anything further in public without violating his own pledge to keep quiet about high-level secrets, so he let the matter rest.

Then came the Snowden revelations, which prompted many to reexamine that exchange. On June 9, the first Sunday after the Snowden leaks were published, Clapper agreed to an interview with NBC-TV's Andrea Mitchell. She asked him why he'd answered Wyden's question the way he did.

Clapper came off as astonishingly unprepared. "I thought, though in retrospect, I was asked 'when are you going to . . . stop beating your wife' kind of question, which is . . . not answered necessarily by a simple yes or no," he began in an incoherent ramble. Then, digging himself still deeper in the hole, he said, "So, I responded in what I thought was the most truthful—or least untruthful—manner by saying, 'No.'"

Doubling down, Clapper homed in on Wyden's use of the word

"collect," as in, "Did the NSA *collect* any type of data . . . on millions of Americans?" Imagine, Clapper said, a vast library of books containing vast amounts of data on every American. "To me," he went on, "*collection* of U.S. persons' data would mean taking the book off the shelf and opening it up and reading it." Therefore, he reasoned, it wasn't quite a lie to say that the NSA did not *collect* data on Americans, at least not wittingly.

The morning after the broadcast, Clapper called his old friend Ken Minihan, the former NSA director, to ask how he did. Minihan was now managing director of the Paladin Capital Group, which invested in cyber security technology firms worldwide: he'd been out of government for more than a decade, but he kept up his contacts throughout the intelligence world; he still had both feet in the game, and he'd watched Clapper's interview in sorrow.

"Well," Minihan replied, in his folksy drawl, "you couldn't have *made* things any worse."

Clapper might have been genuinely perplexed. Five years earlier, the FISA Court had allowed the NSA to redefine "collection" in exactly the way Clapper had done on national television—as the retrieval of data that had already been scooped up and stored. At the time of that ruling, Alexander was laying the foundations of his metadata program; it was illegal to "collect" data from Americans, so the program couldn't have gone forward without redefining the term.

But the FISA Court was a secret body: it met in secret; its cases were heard in secret; its rulings were classified Top Secret. To Clapper and other veterans of the intelligence community, this reworking of a common English word had insinuated its way into official parlance. To anyone outside the walls, the logic seemed disingenuous at best. Clearly, to *collect* meant to gather, to sweep up, to bring together. No one would say, "I'm going to collect *The Great Gatsby* from my bookshelf and read it," nor did it seem plausible that anyone in the

NSA would say, "I'm going to collect this phone conversation from my archive and insert it in my database."

The NSA had basked in total secrecy for so long, from the moment of its inception, that its denizens tended to lose touch with the outside world. In part, its isolation was a product of its mandate: making and breaking codes in the interest of national security ranked among the most sensitive and secretive tasks in all government. Yet the insularity left them without defenses when the bubble was suddenly pierced. They'd had no training or experience in dealing with the public. And as the secrets in Snowden's documents were splashed on front-page headlines and cable newscasts day after jaw-dropping day, the trust in the nation's largest, most intrusive intelligence agency—a trust that had never been more than tenuous—began to crumble.

Opinion polls weren't the only place where the agency took a beating. The lashes were also felt, and more damagingly so, in the agitated statements and angry phone calls from corporate America—in particular, the telecoms and Internet providers, whose networks and servers the NSA had been piggybacking for years, in some cases decades.

This arrangement had been, for many firms, mutually beneficial. As recently as 2009, after the Chinese launched a major cyber attack against Google, stealing the firm's source-code software, the crown jewels of any Internet company, the NSA's Information Assurance Directorate helped repair the damage. One year earlier, after the U.S. Air Force rejected Microsoft's XP operating system on the grounds that it was riddled with security flaws, the directorate helped the firm design XP Service Pack 3, one of the company's most successful systems, which Air Force technicians (and many consumers) deemed secure straight out of the box.

Yet now with their complicity laid bare for all to see, the executives of these corporations backed away, some howling in protest, like Captain Renault, the Vichy official in the film *Casablanca* who pronounced himself "shocked, shocked to find that gambling is going on in here," just as the croupier delivered his winnings for the night. Their fear was that customers in the global marketplace would stop buying their software, suspecting that it was riddled with back doors for NSA intrusion. As Howard Charney, senior vice president of Cisco, a company that had done frequent business with the NSA, told one journalist, the Snowden revelations were "besmirching the reputation of companies of U.S. origin around the world."

Allied governments around the world were clamoring as well. The English-speaking nations that had been sharing intelligence with the United States for decades—the fellow "five-eyes" countries, Great Britain, Canada, Australia, and New Zealand—held firm. But other state leaders, who had not been let into the club, started slipping away. President Obama had planned to rally European leaders in his pressure campaign against China—which had launched cyber attacks on a lot of their companies, too—but his hopes were dashed when a Snowden document revealed that the NSA had once hacked German chancellor Angela Merkel's cell phone. Merkel was outraged.

There was more than a trace of Captain Renault in her fuming, too; as subsequent news stories revealed, the BND, Germany's security service, continued to cooperate with the NSA in monitoring suspected terrorist groups. But at the time, Merkel played populist, as vast swaths of the German people, including many who had once seen America as a protector and friend, started likening the NSA to Stasi, the extremely intrusive surveillance service of the long-imploded East German dictatorship. Other Snowden documents exposed NSA intercepts in Central and South America, infuriating leaders and citizens in the Western Hemisphere, as well.

Something had to be done; the stench—political, economic, and diplomatic—had to be contained. So President Obama did what many of his predecessors had done in the face of crises: he appointed a blue-ribbon commission.

CHAPTER 14

"THE FIVE GUYS REPORT"

O n August 9, 2013, a hot, humid Friday, shortly after three in the afternoon, the laziest hour in the dreariest month for news in the nation's capital, President Obama held a press conference in the East Room of the White House to announce that he was forming "a high-level group of outside experts" to review the charges of abuse in NSA surveillance.

"If you are outside of the intelligence community, if you are the ordinary person and you start seeing a bunch of headlines saying U.S. Big Brother is looking down on you, collecting telephone records, etc., well, understandably," he said, "people would be concerned. I would be concerned too, if I wasn't inside the government." But Obama *was* inside the government, at its apex, and he'd developed a trust in the agencies' propriety. Of course, he acknowledged, "it's not enough for me, as President, to have confidence in these programs. The American people need to have confidence in them as well."

And that, he seemed to be saying, would be the mission of this high-level group of outside experts: not so much to recommend major reforms or even to conduct a particularly hard-hitting probe, but rather, as he put it, to "consider how we can maintain the *trust* of the people." He would also work with Congress to "improve the public's confidence in the oversight conducted by the Foreign Intelligence Surveillance Court." Both efforts would be "designed to ensure that the American people can trust" that the intelligence agencies' actions were "in line with our interests and our values." The high-level group, or the select intelligence committees of Congress, might come up with ways "to jigger slightly" the balance between national security and privacy, and that was fine. "If there are some additional things that we can do to build that trust back up," Obama said, "then we should do them." But he seemed to assume that big changes wouldn't be necessary. "I am comfortable that the program currently is not being abused," he said. "The question is, how do I make the American people more comfortable?"

That same day, as if to seal the case, the Obama administration published a twenty-three-page "white paper," outlining the legal rationale for the bulk collection of metadata from Americans' telephone calls, and the NSA issued its own seven-page memorandum, explaining the program's purpose and constraints.

Already, by this time, Obama, White House chief of staff Denis McDonough, and Susan Rice, his first-term U.N. ambassador who'd recently replaced Tom Donilon as national security adviser, had mulled over possible candidates for the outside group of experts. A few days before the press conference, they chose five, asked them to serve, and, upon getting their consent, ordered the FBI to expedite security-clearance reviews for each.

It wasn't entirely an outside, or independent, group. All five were old friends or former aides of President Obama. Still, it was a more disparate and intriguing bunch than his press conference led many skeptics to expect.

Michael Morell was the establishment pick, a thirty-three-year veteran of the CIA, who had just retired two months earlier as the agency's deputy director and who'd been the main point of contact between Langley and the White House during the secret raid on Osama bin Laden's lair in Pakistan. Morell's presence on the panel would go some distance toward placating the intelligence community.

Two of the choices were colleagues of Obama from his days, in the 1990s, teaching at the University of Chicago Law School. One of them, Cass Sunstein, had also worked on his presidential campaign, served for three years as the chief administrator of his regulatory office, and was married to Samantha Power, his long-standing foreign policy aide, who had recently replaced Susan Rice as U.N. ambassador. An unconventional thinker on issues ranging from the First Amendment to animal rights, Sunstein had written an academic paper in 2008, proposing that government agencies infiltrate the social networks of extremist groups and post messages to undermine their conspiracy theories; some critics of Obama's panel took this paper as a sign that Sunstein was well disposed to NSA domestic surveillance.

The other Chicagoan, Geoffrey Stone, had been dean of the law school when Obama taught there. A prominent member of the ACLU's national advisory council and the author of highly lauded books on the First Amendment in wartime and on excessive secrecy in the national security establishment, Stone seemed a likely critic of NSA abuses.

Peter Swire, a professor of law at the Georgia Institute of Technology, was a longtime proponent of privacy on the Internet and author of a landmark essay on surveillance law. As the White House counsel on privacy during Bill Clinton's presidency, Swire played a key role in the debate over the Clipper Chip, arguing against the NSA's attempt—which he, correctly, saw as futile—to put a clamp on commercial encryption. A couple years later, also on privacy grounds, he argued against Richard Clarke's ill-fated plan to put critical-infrastructure industries on a separate Internet and to wire them so that, in the event of a security breach, the FBI would be directly alerted.

For that reason, Swire was nervous to learn that the fifth member of the Review Group would be Richard Clarke himself. The former White House official who'd immersed himself in NSA practices, written presidential directives on cyber security, and built a reputation as relentless in promoting his own views and in quashing those of others, Clarke was seen as a wild card generally.

Still the consummate operator, Clarke had made a huge splash since quitting the Bush White House on the eve of the Iraq War. One year after the invasion, he gained unlikely fame as an American folk hero at the 9/11 Commission's nationally televised hearings, prefacing his testimony with an apology. "To the loved ones of the victims of 9/11, to them who are here in this room, to those who are watching on television," he began, "your government failed you. Those entrusted with protecting you failed you. And I failed you. We tried hard, but that doesn't matter because we failed. And for that failure, I would ask, once all the facts are out, for your understanding and for your forgiveness."

It seemed to be a genuine plea of contrition—enhanced by the fact that no other Bush official, past or present, had apologized for anything—and the hearing room erupted with applause. After his testimony, family members of victims lined up to thank him, shake his hand, and hug him.

Clarke's critics, whose numbers were legion, scoffed that he was just drumming up publicity. His new book, *Against All Enemies: Inside America's War on Terror*, had hit the bins the previous Friday, trumpeted by a segment on CBS TV's *60 Minutes* the Sunday night between the release date and the hearing. When it soared to the top of the best-seller charts, critics challenged his claims that, in the months leading up to 9/11, Bush's top officials ignored warnings (including Clarke's) of an impending al Qaeda attack and that, the day after the Twin Towers fell, Bush himself pressed Clarke to find evidence pinning the blame on Saddam Hussein to justify the coming war on Iraq. But Clarke, always a scrappy bureaucratic fighter, would never have opened himself to such easy pummeling; he knew the documents would back him up, and, as they trickled to the light of day, they did.

All along, though, Clarke retained his passion for cyber issues, and six years later, he wrote a book called *Cyber War: The Next Threat to National Security and What to Do About It*. Published in April 2010, it was derided by many as overwrought—legitimately in some particulars (he imputed cyber attacks as the possible cause of a few major power outages that had been convincingly diagnosed as freak accidents or maintenance mishaps), but unfairly in the broad scheme of things. Some critics, especially those who knew the author, viewed the book as simply self-aggrandizing: Clarke was now chairman of a cyber-risk-management firm called Good Harbor; thus, they saw *Cyber War* as a propaganda pamphlet to drum up business.

But the main reason for the dismissive response was that the book's scenarios and warnings seemed so unlikely, so sci-fi. The opening of a (generally favorable) review in *The Washington Post* caricatured the skepticism: "Cyber-war, cyber-this, cyber-that: What is it about the word that makes the eyes roll? . . . How authentic can a war be when things don't blow up?"

It had been more than forty years since Willis Ware's paper on the vulnerability of computer networks, nearly thirty years since Ronald Reagan's NSDD-145, and more than a decade since Eligible Receiver, the Marsh Report, Solar Sunrise, and Moonlight Maze—touchstone events in the lives of those immersed in cyberspace, but forgotten, if ever known, to almost everyone else. Even the Aurora Generator Test, just six years earlier, and the offensive cyber operations in Syria, Estonia, South Ossetia, and Iraq—which had taken place more recently still—made little dent on the public consciousness.

Not until a few years *after* Clarke's book—with the revelations about Stuxnet, the Mandiant report on China's Unit 61398, and finally Edward Snowden's massive leak of NSA documents—did cyber espionage and cyber war become the stuff of headline news and everyday conversation. Cyber was suddenly riding high, and when Obama responded to the ruckus by forming a presidential commission, it was only natural that Clarke, the avatar of cyber fright, would be among its appointees.

———

On August 27, the five panelists—christened, that same day, as the President's Review Group on Intelligence and Communications Technologies—met in the White House Situation Room with the president, Susan Rice, and the heads of the intelligence agencies. The session was brief. Obama gave the group's members the deadline for their report—December 15—and assured them that they'd have access to everything they wanted. Three of the panelists were lawyers, so he made it clear that he did not want a legal analysis. Assume that we *can* do this sort of surveillance on legal grounds, he said; your job is to tell me if we *should* be doing it as policy and, if we shouldn't, to come up with something better.

Obama added that he was inclined to follow whatever suggestions they offered, with one caveat: he would not accept any proposal that might impede his ability to stop a terrorist attack.

Through the next four months, the group would meet at least two days a week, sometimes as many as four, often for twelve hours a day or longer, interviewing officials, attending briefings, examining documents, and discussing the implications.

On the first day, shortly before their session with the president, the five met one another, some of them for the first time, in a suite of offices that had been leased for their use. The initial plan had been for them to work inside the national intelligence director's headquarters in Tysons Corner, Virginia, just off the Beltway, ten miles from downtown Washington. But Clarke suggested that they use a more nearby SCIF—a "sensitive compartmented information facility," professionally guarded and structurally shielded to block intruders, electronic and otherwise, from stealing documents or eavesdropping on conversations. Clarke pointed, in particular, to a SCIF on K Street: it would keep the panelists just a few blocks from the White House, and it would preserve their independence, physically and otherwise, from the intelligence community. But Clarke's real motive, which his colleagues realized later, was that this SCIF was located across the street from his consulting firm's office; he preferred not to drive out to the suburbs every day amid the thick rush-hour traffic.

Inside the SCIF that first day, they also met the nine intelligence officers, on loan from various agencies, who would serve as the group's staff. The staffers, one of them explained, would do the administrative work, set the group's appointments, organize its notes, and, at the end, under the group's direction of course, write the report.

The Review Group members looked at one another and smiled; a few laughed. Four of them—Clarke, Stone, Sunstein, and Swire— had written, among them, nearly sixty books, and they had every intention of writing this one, too. This was not going to be the usual presidential commission.

The next morning, they were driven to Fort Meade. Only Clarke and Morell had ever before been inside the place. Clarke's view of the agency was more skeptical than some assumed. In *Cyber War*, he'd criticized the fusion of NSA and Cyber Command under a single four-star general, fearing that the move placed too much power in one person's hands and too much emphasis on cyber offensive operations, at the expense of cyber security for critical infrastructure.

Swire, the Internet privacy scholar, had dealt with NSA officers during the Clipper Chip debate, and he remembered them as smart and professional, but that was fifteen years ago; he didn't know what to expect now. From his study of the FISA Court, he knew about the rulings that let the NSA invoke its foreign intelligence authorities to monitor domestic phone calls; but Edward Snowden's documents, suggesting that the agency was using its powers as an excuse to collect *all calls*, startled him. If this was true, it was way out of line. He was curious to hear the NSA's response.

Stone, the constitutional lawyer and the one member of the group who'd never had contact with the intelligence world, expected to find an agency gone rogue. Stone was no admirer of Snowden: he valued certain whistleblowers who selectively leaked secret information in the interest of the public good; but Snowden's wholesale pilfering of so many documents, of such a highly classified nature, struck him as untenable. Maybe Snowden was right and the government was wrong—he didn't know—but he thought no national security apparatus could function if some junior employee decided which secrets to preserve and which to let fly. Still, the secrets that had come out so far, revealing the vast extent of domestic surveillance, appalled him. Stone had written a prize-winning book about the U.S. government's tendency, throughout history, to overreact in the face of national security threats—from the Sedition Act to the McCarthy era to the surveillance of activists against the Vietnam War—and some of Snowden's documents suggested that the reaction to 9/11 might be another case in point. Stone was already mulling ways to tighten checks and balances.

Upon arrival at Fort Meade, they were taken to a conference room and greeted by a half dozen top NSA officials, including General Alexander and his deputy, John C. "Chris" Inglis. A former Air Force pilot with graduate degrees in computer science, Inglis had spent his entire adult life in the agency, both in its defensive annex and in SIGINT operations; and he'd been among the few dozen extract a list of all the calls that *those* numbers had made and received. But if the analysts wanted to expand the search to a third hop, looking at the numbers called to or from *those* phones, they would have to go through the same procedure all over again, obtaining permission from a supervisor and from the NSA general counsel. (The analysts usually did take a second hop, but almost never a third.)

From the looks that they exchanged across the table, all five members of the Review Group seemed satisfied that the Section 215 program was on the up-and-up (assuming this portion of the briefing was confirmed in a probe of agency files): it was authorized by Congress, approved by the FISA Court, limited in scope, and monitored more fastidiously than any of them had imagined. But President Obama had told them that he didn't want a *legal* opinion of the programs; he wanted a broad judgment of whether they were worthwhile.

So the members asked about the results of this surveillance: How many times had the NSA queried the database, and how many terrorist plots were stopped as a result?

One of the other senior officials had the precise numbers at hand. For all of 2012, the NSA queried the database for 288 U.S. phone numbers. As a result of those queries, the agency passed on twelve "tips" to the FBI. If the FBI found the tips intriguing, *it* could request a court order to intercept the calls to and from that phone number— to *listen in* on the calls—using NSA technology, if necessary.

So, one of the commissioners asked, how many of those twelve tips led to the halting of a plot or the capture of a terrorist?

The answer was zero. None of the tips had led to anything worth pursuing further; none of the suspicions had panned out.

Geof Stone was floored. "Uh, *hello*?" he thought. "What are we *doing* here?" The much-vaunted metadata program (a) seemed to be tightly controlled, (b) did *not* track every phone call in America, and, now it turned out, (c) had not unearthed a single terrorist.

bright young men that Ken Minihan and Mike Hayden promote
ahead of schedule as part of the agency's post–Cold War reforms.

After some opening remarks, Alexander made his exit, returning
periodically through the day, leaving Inglis in charge. Over the next
five hours, Inglis and the other officials gave rotating briefings on the
controversial surveillance programs, delving deeply into the details.

The most controversial program was the bulk collection of telephone metadata, as authorized by Section 215 of the Patriot Act. According to the Snowden documents, this allowed the NSA to collect
and store the records of *all* phone calls inside the United States—not
the contents of those calls, but the phone numbers of those talking,
as well as the dates, times, and durations of the conversations, which
could reveal quite a lot of information on their own.

Inglis told the group that, in fact, this was not how the program
really operated. In the FISA Court's ruling on Section 215, the NSA
could delve into this metadata, looking for connections among various phone numbers, *only* for the purpose of finding associates of
three specific foreign terrorist organizations, including al Qaeda.

Clarke interrupted him. You've gone to all the trouble of setting
up this program, he said, and you're looking for connections to *just
three organizations?*

That's all we have the authority to do, Inglis replied. Moreover,
if the metadata revealed that someone inside the United States had
called, or been called by, a suspected terrorist, just twenty-two people in the entire NSA—twenty line personnel and two supervisors—
were able to request and examine more data about that phone number.
And before that data could be probed, two of those twenty personnel
and at least one of the supervisors had to agree, independently, that
an expanded search was worthwhile. Finally, the authority to search
that person's phone records would expire after 180 days.

If something suspicious showed up in one of those numbers, the
NSA analysts could take a second hop; in other words, they could

Clarke asked the unspoken question: Why do you still *have* this program if it hasn't produced any results?

Inglis replied that the program had hastened the speed with which the FBI captured at least one terrorist. And, he added, it might point toward a plot sometime in the future. The metadata, after all, exist; the phone companies collect it routinely, as "business records," and would continue to do so, with or without the NSA or Section 215. Since it's there, why not use it? If someone in the United States phoned a known terrorist, wasn't it *possible* that a plot was in the works? As long as proper safeguards were taken to protect Americans' privacy, why *not* look into it?

The skeptics remained tentatively unconvinced. This was something to examine more deeply.

Inglis moved on to what he and his colleagues considered a far more important and damaging Snowden leak. It concerned the program known as PRISM, in which the NSA and FBI tapped into the central servers of nine leading American Internet companies—mainly Microsoft, Yahoo, and Google, but also Facebook, AOL, Skype, YouTube, Apple, and Paltalk—extracting email, documents, photos, audio and video files, and connection logs. The news stories about PRISM acknowledged that the purpose of the intercepts was to track down exclusively foreign targets, but the stories also noted that ordinary Americans' emails and cellular phone calls got scooped up in the process as well.

The NSA had released a statement, right after the first news stories, calling PRISM "the most significant tool in the NSA's arsenal for the detection, identification, and disruption of terrorist threats to the US and around the world." General Alexander had publicly claimed that the data gathered from PRISM had helped discover and disrupt the planning of fifty-four terrorist attacks—a claim that Inglis now repeated, offering to share all the case files with the Review Group.

Whatever the ambiguities about the telephone metadata program, he stated, PRISM had demonstrably saved lives.

Did Americans' calls and email get caught up in the sweep? Yes, but that was an unavoidable by-product of the technology. The NSA briefers explained to the Review Group what Mike McConnell had explained, back in 2007, to anyone who'd listen: that digital communications traveled in packets, flowing along the most efficient path; and, because most of the world's bandwidth was concentrated in the United States, pieces of almost every email and cell phone conversation in the world flowed, at some point, through a line of American-based fiber optics.

In the age of landlines and microwave transmissions, if a terrorist in Pakistan called a terrorist in Yemen, the NSA could intercept their conversation without restraint; now, though, if the same two people, in the same overseas locations, were talking on a cell phone, and if NSA analysts wanted to latch on to a packet containing a piece of that conversation while it flowed inside the United States, they would have to get a warrant from the Foreign Intelligence Surveillance Court. It made no sense.

That's why McConnell pushed for a revision in the law, and that's what led to the Protect America Act of 2007 and to the FISA Amended Act of 2008, especially Section 702, which allowed the government to conduct electronic surveillance inside the United States—"with the assistance of a communications service provider," in the words of that law—as long as the people communicating were "reasonably believed" to be outside the United States.

The nine Internet companies, which were named in the news stories, had either complied with NSA requests to tap into their servers or been ordered by the FISA Court to let the NSA in. Either way, the companies had long known what was going on.

Much of this was clear to the Review Group, but some of the procedures that Inglis and the others described were baffling. What did it mean that callers were "reasonably believed" to be on foreign soil? How did the NSA analysts make that assessment?

The briefers went through a list of "selectors"—key-word searches and other signposts—that indicated possible "foreignness." As more selectors were checked off, the likelihood increased. The intercept could legally get under way, once there was a 52 percent probability that both parties to the call or the email were foreign-based.

Some on the Review Group commented that this seemed an iffy calculation and that, in any case, 52 percent marked a very low bar. The briefers conceded the point. Therefore, they went on, if it turned out, once the intercept began, that the parties were *inside* the United States, the operation had be shut down immediately and all the data thus far retrieved had to be destroyed.

The briefers also noted that, even though a court order wasn't required for these Section 702 intercepts, the NSA couldn't go hunting for just anything. Each year, the agency's director and the U.S. attorney general had to certify, in a list approved by the FISA Court, the *categories* of intelligence targets that could be intercepted under Section 702. Then, every fifteen days, after the start of a new intercept, a special panel inside the Justice Department reviewed the operation, making sure it conformed to that list. Finally, every six months, the attorney general reviewed all the start-ups and submitted them to the congressional intelligence committees.

But there was a problem in all this. To get at the surveillance target, the NSA operators had to scoop up the entire *packet* that carried the pertinent communication. This packet was interwoven with other packets, which carried pieces of other communications, many of them no doubt involving Americans. What happened to all of those pieces? How did the agency make sure that some analyst didn't read those emails or listen to those cell phone conversations?

The briefers raised these questions on their own, because, just one week earlier, President Obama had declassified a ruling, back in October 2011, by a FISA Court judge named John Bates, excoriating the NSA for the Section 702 intercepts generally. The fact that domestic communications were caught up in these "upstream collections," as they were called, was no accident, Bates wrote in his ruling; it was an inherent part of the program, an inherent part of packet-switching technology. Unavoidably, then, the NSA was collecting "tens of thousands of wholly domestic communications" each year, and, as such, this constituted a blatant violation of the Fourth Amendment.

"The government," Bates concluded, "has failed to demonstrate that it has struck a reasonable balance between its foreign intelligence needs and the requirement that information concerning U.S. persons be protected." As a result, he ordered a shutdown of the entire Section 702 program, until the NSA devised a remedy that did strike such a balance, and he ordered the agency to delete all upstream files that had been collected to date.

This was a serious legal problem, the briefers acknowledged, but, they emphasized, it had been brought to the court's attention *by* the NSA; there was no cover-up of wrongdoing. After Bates's ruling, the NSA changed the architecture of the collection system in a way that would minimize future violations. The new system was put in place a month before the Review Group was formed; Judge Bates declared himself satisfied that it solved the problem.

All in all, the first day of the Review Group's work was productive. The NSA officials around the table had answered every question, taken up every challenge with what seemed to be genuine candor, even an interest in discussing the issues. They'd rarely discussed these matters with outsiders—until then, no outsider had been cleared to discuss them—and they seemed to relish the opportunity. Geoffrey Stone in particular was impressed; the tenor seemed more like a university seminar than a briefing inside the most cloistered American intelligence agency.

It also seemed clear—if the officials were telling the truth (an assumption the Review Group would soon examine)—that, in one sense, the Snowden documents had been overblown. Stone's premise going into the day—that the NSA had morphed into a rogue agency—seemed invalid: the programs that Snowden uncovered (again, assuming the briefings were accurate) had been authorized, approved, and pretty closely monitored. Most of the checks and balances that Stone had thought about proposing, it turned out, were already in place.

But to some of the panelists, certainly to Stone, Swire, and Clarke, the briefings had not dispelled a larger set of concerns that the Snowden leaks had raised. These NSA officials, who'd been briefing them all day long, seemed like decent people; the safeguards put in place, the standards of restraint, were impressive; clearly, this was like neither the NSA of the 1960s nor an intelligence agency of any other country. But what if the United States experienced a few more terrorist attacks? Or what if a different sort of president, or a truly roguish NSA director, came to power? Those restraints had been put up from the inside, and they could be taken down from the inside as well. Clearly, the agency's technical prowess was staggering: its analysts could penetrate every network, server, phone call, or email they wished. The law might bar them from looking at, or listening to, the contents of those exchanges, but if the law were changed or ignored, there would be no physical obstacles; if the software were reprogrammed to track down political dissidents instead of terrorists, there would be no problem compiling vast databases on those kinds of targets.

In short, there was enormous *potential* for abuse. Stone, who'd written a book about the suppression of dissent in American history, shivered at the thought of what President Richard Nixon or FBI director J. Edgar Hoover might have done if they'd had this technology at their fingertips. And who could say, especially in the age of terror, that Americans would never again see the likes of Nixon or Hoover in the upper echelons of power?

———

Stone nurtured an unexpected convert to this view in Mike Morell, the recently retired spy on the Review Group. The two shared an office in the SCIF on K Street, and Stone, a charismatic lecturer, laid out the many paths to potential abuse as well as the incidents of actual abuse in recent times, a history of which Morell claimed he knew little, despite his three decades in the CIA. (During the Church hearings, Morell was in high school, oblivious to global affairs; his posture at Langley, where he went to work straight out of college, was that of a nose-to-the-grindstone Company Man.)

Over the next four months, the group returned to Fort Meade a few times, and delegations from Fort Meade visited the group at its office a few times, as well. The more files that the group and its staff examined, the more they felt confirmed in their impressions from the first briefing.

Morell was the one who pored through the NSA case files, including the raw data from all fifty-four terrorist plots that Alexander and Inglis claimed were derailed because of the PRISM program under Section 702 of the FISA Act, as well as a few plots that they were now claiming, belatedly, had been unearthed because of the bulk collection of telephone metadata, authorized by Section 215 of the Patriot Act. Morell and the staff, who also reviewed the files, concluded that the PRISM intercepts did play a role in halting fifty-three of those fifty-four plots—a remarkable validation of the NSA's central counterterrorist program. However, in *none* of those fifty-three files did they find evidence that metadata played a substantial role. Nor were they persuaded by the few new cases that Alexander had sent to the group: yes, in those cases, a terrorist's phone number showed up in the metadata, but it showed up in several other intercepts, too. Had there never been a Section 215, had metadata never been collected in bulk, the NSA or the FBI would still have uncovered those plots.

This conclusion came as a surprise. Morell was inclined to assume that highly touted intelligence programs produced results. His findings to the contrary led Clarke, Stone, and Swire to recommend killing the metadata program outright. Morell wasn't prepared to go that far; neither was Sunstein. Both bought the argument that, even if it hadn't stopped any plots so far, it might do so in the future. Morell went further still, arguing that the absence of results suggested that the program should be intensified. For a while, the group's members thought they'd have to issue a split verdict on this issue.

Then, during one of the meetings at Fort Meade, General Alexander told the group that he could live with an arrangement in which the telecom companies held on to the metadata and the NSA could gain access to specified bits of it only through a FISA Court order. It might take a little longer to obtain the data, but not by much, a few hours maybe; and the new rule, Alexander suggested, could provide for exceptions, allowing access, with a post-facto court order, in case of an emergency.

Alexander also revealed that the NSA once had an *Internet* metadata program, but it proved very expensive and yielded no results, so, in 2011, he'd terminated it.

To the skeptics on the Review Group, this bit of news deepened their suspicions about Section 215 and the worth of metadata generally. The NSA had a multibillion-dollar budget; if Internet metadata had produced *anything* of promise, Alexander could have spent more money to expand its reach. The fact that he didn't, the fact that he killed the program rather than double down on the investment, splashed doubts on the concept's value.

Even Morell and Sunstein seemed to soften their positions: if Alexander was fine with storing metadata outside the NSA, that might be a compromise the entire group could get behind. Morell embraced the notion with particular enthusiasm: metadata would still exist; but removing it from NSA headquarters would prevent some future rogue director from examining the data at will—would minimize the potential for abuse that Stone had convinced him was a serious issue.

The brief dispute over metadata—whether to kill the program or expand it—had sparked one of the few fits of rancor within the group. This fact had been another source of surprise: given their disparate backgrounds and beliefs, the members had expected to be at each other's throats routinely. From early on, though, the atmosphere was harmonious.

This camaraderie took hold on the second day of their work when the five went to the J. Edgar Hoover Building—FBI headquarters—in downtown Washington. The group's staff had requested detailed briefings on the bureau's relationship with the NSA and on its own version of metadata collection, known as National Security Letters, which, under Section 505 of the Patriot Act, allowed access to Americans' phone records and other transactions that were deemed "relevant" to investigations into terrorism or clandestine intelligence activities. Unlike the NSA's metadata program, the FBI's had no restrictions at all: the letters required no court order; any field officer could issue one, with the director's authorization; and the recipients of a letter were prohibited from *ever* revealing that they'd received it. (Until a 2006 revision, they couldn't even inform their lawyer.) Not merely the potential for abuse, but actual instances of abuses, seemed very likely.

When the five arrived at the bureau's headquarters, they were met not by the director, nor by his deputy, but by the third-ranking official, who took leave after escorting them to a conference room, where twenty FBI officials sat around a table, prepared to drone through canned presentations, describing their jobs, one by one, for the hour that the group had been allotted.

Ten minutes into this dog-and-pony show, Clarke asked about

the briefings that they'd requested. Specifically, he wanted to know how many National Security Letters the FBI issued each year and how the bureau measured their effectiveness. One of the officials responded that only the regional bureaus had those numbers, no one had collated them nationally; and no one had devised any measure of effectiveness.

The canned briefings resumed, but after a few more minutes, Clarke stood up and exclaimed, "This is bullshit. We're out of here." He walked out of the room; the other four sheepishly followed, while the FBI officials sat in shock. At first, Clarke's colleagues were a bit mortified, too; they'd heard about his antics and wondered if this was going to be standard procedure.

But by the next day, it was clear that Clarke had known exactly what he was doing: word quickly spread about "the bullshit briefing," and from that point on, no federal agency dared to insult the group with condescending show-and-tell; only a few agencies proved to be very useful, but they all at least tried to be substantive, and the FBI even called back for a second chance.

Clarke's act emboldened his colleagues to press more firmly for answers to their questions. The nature of their work reinforced this solidarity. They were the first group of outsiders to investigate this subject with the president's backing, and they derived an esprit de corps from the distinction. More than that, they found themselves agreeing on almost everything, because most of the facts seemed so clear. Peter Swire, who'd been nervous about reigniting fifteen-year-old tensions with Clarke, found himself getting along grandly with his former rival and gaining confidence in his own judgments, the more they aligned with his.

As the air lightened from cordiality to jollity, they started calling themselves the "five guys," after the name of a local hamburger joint, and referring to the big book they'd soon be writing as "The Five Guys Report."

Their esprit intensified with the realization that they were pretty much the only serious monitors in town. They met on Capitol Hill with the select intelligence committees, and came away concluding that their members had neither the time nor the resources for deep oversight. They spoke with a few former FISA Court judges and found them too conciliatory by nature.

Good thing, they concluded, that the NSA had an inside army of lawyers to assure compliance with the rules, because if it didn't, no one on the outside would have any way of knowing whether the agency was or wasn't a den of lawlessness. The five guys agreed that their task, above all else, was to come up with ways to strengthen external controls.

They divvied up the writing chores, each drafting a section or two and inserting ideas on how to fix the problems they'd diagnosed. Cut, pasted, and edited, the sections added up to a 303-page report, with forty-six recommendations for reform.

One of the key recommendations grew out of the group's conversation with General Alexander: all metadata should be removed from Fort Meade and held by the private telecom companies or some other third party, with the NSA allowed access only through a FISA Court order. The group was particularly firm on this point. Even Mike Morell had come to view this recommendation as the report's centerpiece: if the president rejected it, he felt, the whole exercise will have been pointless.

Another proposal was to bar the FBI from issuing National Security Letters without a FISA Court order and, in any case, letting recipients disclose that they'd received such a letter after 180 days, unless a judge extended the term of secrecy for specific national security reasons. The point of both recommendations was, as the report put it, to "reduce the risk, both actual and perceived, of government abuse."

The group also wrote that the FISA Court should include a public interest advocate, that NSA directors should be confirmed by the Senate, that they should not take on the additional post of U.S. cyber commander (on the grounds that dual-heading CyberCom and the NSA gave too much power to one person), and that the Information Assurance Directorate—the cyber security side of Fort Meade—should be split off from the NSA and turned into a separate agency of the Defense Department.

Another recommendation was to bar the government from doing anything to "subvert, undermine, weaken, or make vulnerable generally available commercial software." Specifically, if NSA analysts discovered a zero-day exploit—a vulnerability that no one had yet discovered—they should be required to patch the hole at once, except in "rare instances," when the government could "briefly authorize" using zero-days "for high-priority intelligence collection," though, even then, they could do so only after approval by a "senior interagency review involving all appropriate departments."

This was one of the group's more esoteric, but also radical, recommendations. Zero-day vulnerabilities were the gemstones of modern SIGINT, prized commodities that the agency trained its top sleuths—and sometimes paid private hackers—to unearth and exploit. The proposal was meant, in part, to placate American software executives, who worried that the foreign market would dry up if prospective customers assumed the NSA had carved back doors into their products. But it was also intended to make computer networks less vulnerable; it was a declaration that the needs of cyber security should supersede those of cyber offensive warfare.

Finally, lest anyone interpret the report as an apologia for Edward Snowden (whose name appeared nowhere in the text), ten of the forty-six recommendations dealt with ways to tighten the security of highly classified information inside the intelligence community, including procedures to prevent system administrators—which had been Snowden's job at the NSA facility in Oahu—from gaining access to documents unrelated to their work.

It was a wide-ranging set of proposals. Now what to do with them? When the five had first met, back in late August, they'd started to discuss how to handle the many disagreements that they anticipated. Should the report contain dissenting footnotes, majority and minority chapters, or what? They'd never finished that conversation, and now, here they were, with the mid-December deadline looming.

One of the staffers suggested listing the forty-six recommendations on an Excel spreadsheet, with the letters Y and N (for Yes and No) beside each one, and giving a copy of the sheet to each of the five members, who would mark it up, indicating whether they agreed or disagreed with each recommendation. The staffer would then tabulate the results.

After counting the votes, the staffer looked up and said, "You won't believe this." The five guys had agreed, unanimously, on all forty-six.

———

On December 13, two days before deadline, the members of the Review Group turned in their report, titled *Liberty and Security in a Changing World*. To their minds, they fulfilled their main task—as their report put it, "to promote public trust, while also allowing the Intelligence Community to do what must be done to respond to genuine threats"—but also exceeded that narrow mandate, outlining truly substantial reforms to the intelligence collection system.

Their language was forthright in ways bound to irritate all sides of the debate, which had only intensified in the six months since the onslaught of Snowden-leaked documents. "Although recent disclosures and commentary have created the impression, in some quarters, that NSA surveillance is indiscriminate and pervasive across the globe," the report stated, "that is not the case." However, it went on, the Review Group did find "serious and persistent instances of noncompliance in the Intelligence Community's implementation of its authorities," which, "even if unintentional," raised "serious concerns" about its "capacity to manage" its powers "in an effective, lawful manner."

To put it another way (and the point was made several times, throughout the report), while the group found "no evidence of illegality or other abuse of authority for the purpose of targeting domestic political activities," there was, always present, "the lurking danger of abuse." In a passage that might have come straight out of Geoffrey Stone's book, the report stated, "We cannot discount the risk, in light of the lessons of our own history, that at some point in the future, high-level government officials will decide that this massive database of extraordinarily sensitive private information is there for the plucking."

On December 18, at eleven a.m., President Obama met with the group again in the Situation Room. He'd read the outlines of the report and planned to peruse it during Christmas break at his vacation home in Hawaii.

A month later, on January 17, 2014, in a speech at the Justice Department, Obama announced a set of new policies prompted by the report. The first half of his speech dwelled on the importance of intelligence throughout American history: Paul Revere's ride, warning that the British were coming; reconnaissance balloons floated by the Union army to gauge the size of Confederate regiments; the vital role of code-breakers in defeating Nazi Germany and Imperial Japan. Similarly, today, he said, "We cannot prevent terrorist attacks or cyber threats without some capability to penetrate digital communications."

The message had percolated throughout the national security bureaucracy, and Obama had absorbed it as well: that, in the cyber world, offense and defense stemmed from the same tools and techniques. (In a interview several months later with an IT webzine, Obama, the renown hoops enthusiast, likened cyber conflict to basketball, "in the sense that there's no clear line between offense and defense, things are going back and forth all the time.") And so, the president ignored the Review Group's proposals to split the NSA from Cyber Command or to place the defensive side of NSA in a separate agency.

However, Obama agreed with the group's general point on "the risk of government overreach" and the "potential for abuse." And so, he accepted many of its other recommendations. He rejected the proposal to require FISA Court orders for the FBI's National Security Letters, but he did limit how long the letters could remain secret. (He eventually settled on a limit of 180 days, with a court order required for an extension.) There would be no more surveillance of "our close friends and allies" without some compelling reason (a reference to the monitoring of Angela Merkel's cell phone, though Obama's wording allowed a wide berth for exceptions). And his national security team would conduct annual reviews of the surveillance programs, weighing security needs against policies toward alliances, privacy rights, civil liberties, and the commercial interests of U.S. companies.

This last idea led, three months later, to a new White House policy barring the use of a zero-day exploit, unless the NSA made a compelling case that the pros outweighed the cons. And the final verdict on its case would be decided not by the NSA director but by the cabinet secretaries in the NSC and, ultimately, by the president.

This was potentially a very big deal. Whether it would really limit the practice—whether it amounted to a political check or a rubber stamp—was another matter.*

Finally, Obama spoke of the most controversial program, the bulk collection of telephone metadata under Section 215 of the Patriot Act. First, as an immediate step, he ordered the NSA to restrict its data searches to two hops, down from its previously allowed limit of three. (Though potentially significant, this had little real impact, as the NSA almost never took three hops.) Second, and more significant, he endorsed the proposal to *store* the metadata with a private entity and to allow NSA access only after a FISA Court order.

These endorsements seemed doomed, though, because any changes in the storage of metadata or in the composition of the FISA Court would have to be voted on by Congress. Under ordinary conditions, Congress—especially this Republican-controlled Congress—wouldn't schedule such a vote: its leaders had no desire to change the operations of the intelligence agencies or to do much of anything that President Obama wanted them to do.

But these weren't ordinary conditions. The USA Patriot Act had been passed by Congress, under great pressure, in the immediate aftermath of the September 11 attacks: the bill came to the floor hot off the printing presses; almost no one had time to read it. In exchange for their haste in passing it, key Democratic legislators insisted, over intense opposition by the Bush White House, that a sun-

---

* The questions to be asked, in considering whether to exploit a zero-day vulnerability, were these: To what extent is the vulnerable system used in the critical infrastructure; in other words, does the vulnerability, if left unpatched, pose significant risk to our own society? If an adversary or criminal group knew about the vulnerability, how much harm could it inflict? How likely is it that we would know if someone else exploited it? How badly do we need the intelligence we think we can get from exploiting it? Are there other ways to get the intelligence? Could we exploit the vulnerability for a short period of time before disclosing and patching it?

set clause—an expiration date—be written into certain parts of the law (including Section 215, which allowed the NSA to collect and store metadata), so that Congress could extend its provisions, or let them lapse, at a time allowing more deliberation.

In 2011, when those provisions had last been set to expire, Congress voted to extend them until June 2015. In the interim four years, three things happened. First, and pivotally, came Edward Snowden's disclosures about the extent of NSA domestic surveillance. Second, the five guys report concluded that this metadata hadn't nabbed a single terrorist and recommended several reforms to reduce the potential for abuse.

Third, on May 7, just weeks before the next expiration date, the U.S. 2nd Circuit Court of Appeals ruled that Section 215 of the Patriot Act did not in fact authorize anything so broad as the NSA's bulk metadata collection program—that the program was, in fact, illegal. Section 215 permitted the government to intercept and store data that had "relevance" to an "investigation" of a terrorist plot or group. The NSA reasoned that, in tracing the links of a terrorist conspiracy, it was impossible to know what was relevant—who the actors were—ahead of time, so it was best to create an archive of calls that could be plowed through in retrospect; it was necessary, by this logic, to collect *everything* because anything *might* prove relevant; to find a needle in a haystack, you needed access to "the whole haystack." The FISA Court had long ago accepted the NSA's logic, but now the 2nd Circuit Court rejected it as "unprecedented and unwarranted." In the court case that culminated in the ruling, the Justice Department (which was defending the NSA position) likened the metadata collection program to the broad subpoena powers of a grand jury. But the court jeered at the analogy: grand juries, it noted, are "bounded by the facts" of a particular investigation and "by a finite time limitation," whereas the NSA metadata program required "that the phone companies turn over records on an 'ongoing daily basis'—with no foreseeable end point, no requirement of relevance to any particular set of facts, and no limitations as to subject matter or individuals covered."

The judges declined to rule on the program's constitutionality; they even allowed that Congress could authorize the metadata program, if it chose to do so explicitly. And so it was up to Congress—and its members couldn't evade the moment of truth. Owing to the sunset clause, the House and Senate *had* to take a vote on Section 215, one way or the other; if they didn't, the metadata program would expire by default.

In this altered climate, the Republican leaders couldn't muster majority support to sustain the status quo. Moderates in Congress drafted a bill called the USA Freedom Act, which would keep metadata stored with the telecom companies and allow the NSA access only to narrowly specified pieces of it, and only after obtaining a FISA Court order to do so. The new law would also require the FISA Court to appoint a civil-liberties advocate to argue, on occasion, against NSA requests; and it would require periodic reviews to declassify at least portions of FISA Court rulings. The House passed the reform bill by a wide majority; the Senate, after much resistance by the Republican leadership, had no choice but to pass it as well.

Against all odds, owing to the one bit of farsighted caution in a law passed in 2001 amid the panic of a national emergency, Congress approved the main reforms of NSA practices, as recommended by President Obama's commission—and by President Obama himself.

The measures wouldn't change much about cyber espionage, cyber war, or the long reach of the NSA, to say nothing of its foreign counterparts. For all the political storms that it stirred, the bulk collection of domestic metadata comprised a tiny portion of the agency's activities. But the reforms would block a tempting path to potential abuse, and they added an extra layer of control, albeit a thin one, on the agency's power—and its technologies' inclination—to intrude into everyday life.

On March 31, two and a half months after Obama's speech at the Justice Department, in which he called for those reforms, Geoffrey Stone delivered a speech at Fort Meade. The NSA staff had asked him to recount his work on the Review Group and to reflect on the ideas and lessons he'd taken away.

Stone started off by noting that, as a civil libertarian, he'd approached the NSA with great skepticism, but was quickly impressed by its "high degree of integrity" and "deep commitment to the rule of law." The agency made mistakes, of course, but they were just that—mistakes, not intentional acts of illegality. It wasn't a rogue agency; it was doing what its political masters wanted and what the courts allowed, and, while reforms were necessary, its activities were generally lawful.

His speech lavished praise a little while longer on the agency and its employees, but then it took a sharp turn. "To be clear," he emphasized, "I am not saying that citizens should *trust* the NSA." The agency needed to be held up to "constant and rigorous review." Its work was "important to the safety of the nation," but, by nature, it posed "grave dangers" to American values.

"I found, to my surprise, that the NSA deserves the respect and appreciation of the American people," he summed up. "But it should never, ever, be trusted."

CHAPTER 15

"WE'RE WANDERING IN
DARK TERRITORY"

I N the wee hours of Monday, February 10, 2014, four weeks after
President Obama's speech at the Justice Department on NSA re-
form, hackers launched a massive cyber attack against the Las Vegas
Sands Corporation, owner of the Venetian and Palazzo hotel-casi-
nos on the Vegas Strip and a sister resort, the Sands, in Bethlehem,
Pennsylvania.

The assault destroyed the hard drives in thousands of servers,
PCs, and laptops, though not before stealing thousands of customers'
credit-card charges as well as the names and Social Security numbers
of company employees.

Cyber specialists traced the attack to the Islamic Republic of Iran.

The previous October, Sheldon Adelson, the ardently pro-Israel,
right-wing billionaire who owned 52 percent of Las Vegas Sands
stock, had spoken on a panel at Yeshiva University in New York. At
one point, he was asked about the Obama administration's ongoing
nuclear negotiations with Iran.

"What I would say," he replied, "is, 'Listen. You see that desert out there? I want to show you something.'" Then, Adelson said, he would drop a nuclear bomb on the spot. The blast "doesn't hurt a soul," he went on, "maybe a couple of rattlesnakes or a scorpion or whatever." But it does lay down a warning: "You want to be wiped out?" he said he'd tell the mullahs. "Go ahead and take a tough position" at those talks.

Adelson's monologue went viral on YouTube. Two weeks later, the Ayatollah Ali Khamenei, Iran's supreme leader, fumed that America "should slap these prating people" and "crush their mouths."

Soon after, the hackers went to work on Adelson's company. On January 8, they tried to break into the Sands Bethlehem server, probing the perimeters for weak spots. On the twenty-first, and again on the twenty-sixth, they activated password-cracking software, trying out millions of letter-and-number combinations, almost instantaneously, to hack into the company's Virtual Private Network, which employees used at home or on the road.

Finally, on February 1, they found a weakness in the server of a Bethlehem company that tested new pages for the casino's website. Using a tool called Mimikatz, which extracted all of a server's recent records, the hackers found the login and password of a Sands systems engineer who'd just been in Bethlehem on a business trip. Using his credentials, they strolled into the Vegas-based servers, probed their pathways, and inserted a malware program, consisting of just 150 lines of code, that wiped out the data stored on every computer and server, then filled the spaces with a random stream of zeroes and ones, to make restoring the data nearly impossible.

Then they started to download really sensitive data: the IT passwords and encryption keys, which could take them into the mainframe computer, and, potentially more damaging, the files on high-rolling customers—"the whales," as casino owners called them.

Just in time, Sands executives shut off the company's link to the Internet.

Still, the next day, the hackers found another way back in and defaced the company's website with a message: "Encouraging the Use of Weapons of Mass Destruction UNDER ANY CONDITION Is a Crime." Then they shut down a few hundred more computers that hadn't been disabled the first time around.

After the storm passed, the casino's cyber security staff estimated that the Iranians had destroyed twenty thousand computers, which would cost at least $40 million to replace.

It was a typical, if somewhat sophisticated, cyber attack for the second decade of the twenty-first century. Yet there was one thing odd about these hackers: anyone breaking into the servers of a Las Vegas resort hotel casino could have made off with deep pools of cash—but these hackers didn't take a dime. Their sole aim was to punish Sheldon Adelson for his crude comments about nuking Iran: they launched a cyber attack not to steal money or state secrets, but to influence a powerful man's political speech.

It was a new dimension, a new era, of cyber warfare.

Another notable feature, which the Sands executives picked up on after the fact: the Iranians were able to unleash such a destructive attack, after making such extensive preparations, without arousing notice, because the company's cyber security staff consisted of just five people.

Las Vegas Sands—one of the largest resort conglomerates in the world, with forty thousand employees and assets exceeding $20 billion—wasn't ready to deal with the old era of cyber war, much less the new one.

At first, not wanting to scare off customers, the executives tried to cover up just how badly the hack had hurt them, issuing a press release commenting only on their website's defacement. The hackers struck back, posting a video on YouTube showing a computer screen with what seemed like thousands of the Sands' files and folders, including passwords and casino credit records, underscored with a text box reading, "Do you really think that only your mail server has been taken down?!! Like hell it has!!"

The FBI took down the video within a few hours, and the company managed to quash much further exposure, until close to the end of the year, when *Bloomberg Businessweek* published a long story detailing the full scope of the attack and its damage. But the piece drew little notice because, two weeks earlier, a similar, though far more devastating attack hit the publicity-drenched world of Hollywood, specifically one of its major studios—Sony Pictures Entertainment.

On Monday morning, November 24, a gang of hackers calling themselves "Guardians of Peace" hacked into Sony Pictures' network, destroying three thousand computers and eight hundred servers, carting off more than one hundred terabytes of data—much of which was soon sent to, and gleefully reprinted by, the tabloid, then the mainstream, press—including executives' salaries, emails, digital copies of unreleased films, and the Social Security numbers of 47,000 actors, contractors, and employees.

Sony had been hacked before, twice in 2011 alone: one of the attacks shut down its PlayStation network for twenty-three days after purloining data from 77 million accounts; the other stole data from 25 million viewers of Sony Online Entertainment, including twelve thousand credit card numbers. The cost, in business lost and damages repaired, came to about $170 million.

But, like many conglomerates, Sony ran its various branches in stovepipe fashion: the executives at PlayStation had no contact with those at Online Entertainment, who had no contact with those at Sony Pictures. So the lessons learned in one realm were not shared with the others.

Now, the executives realized, they had to get serious. To help track down the hacker and fix the damage, they contacted not only the FBI but also FireEye, which had recently purchased Mandiant, the company—headed by the former Air Force cyber crime investigator Kevin Mandia—that had, most famously, uncovered the massive array of cyber attacks launched by Unit 61398 of the Chinese army. Soon enough, both FireEye and the FBI, the latter working with NSA, identified the attackers as a group called "DarkSeoul," which often did cyber jobs for the North Korean government from outposts scattered across Asia.

Sony Pictures had planned to release on Christmas Day a comedy called *The Interview*, starring James Franco and Seth Rogen as a frothy TV talk show host and his producer who get mixed up in a CIA plot to assassinate North Korea's ruler, Kim Jong-un. The previous June, when the project was announced, the North Korean government released a statement warning that it would "mercilessly destroy anyone who dares hurt or attack the supreme leadership of the country, even a bit." The hack, it seemed, was the follow-up to the threat.

Some independent cyber specialists doubted that North Korea was behind the attack, but those deep inside the U.S. intelligence community were unusually confident. In public, officials said that the hackers used many of the same "signatures" that DarkSeoul had used in the past, including an attack two years earlier that wiped out forty thousand computers in South Korea—the same lines of code, encryption algorithms, data-deletion methods, and IP addresses. But the real reason for the government's certainty was that the NSA had long ago penetrated North Korea's networks: anything that its hackers did, the NSA could follow; when the hackers monitored what they were doing, the NSA could intercept the signal from their monitors—not in real time (unless there was a reason to be watching the North Koreans in real time), but the agency's analysts could retrieve the files, watch the images, and compile the evidence retroactively.

It was another case of a cyber attack launched not for money, trade secrets, or traditional espionage, but to influence a private company's behavior.

This time, the blackmail worked. One week before opening day, Sony received an email threatening violence against theaters showing the film. Sony canceled its release; and, suddenly the flow of embarrassing emails and data to the tabloids and the blogosphere ceased.

The studio's cave-in only deepened its problems. At his year-end press conference, traditionally held just before flying off to his Hawaii home for the holidays, President Obama told the world that Sony "made a mistake" when it canceled the movie. "I wish they had spoken to me first," he went on. "I would have told them, 'Do not get into a pattern in which you're intimidated by these kinds of criminal acts.'" He also announced that the United States government would "respond proportionally" to the North Korean attack, "in a place and time and manner that we choose."

Some in the cyber world were perplexed. Hundreds of American banks, retailers, utilities, defense contractors, even Defense Department networks had been hacked routinely, sometimes at great cost, with no retributive action by the U.S. government, at least not publicly. But a Hollywood studio gets breached, over a *movie*, and the president pledges retaliation in a televised news conference?

Obama did have a point in making the distinction. Jeh Johnson, the secretary of homeland security, said on the same day that the Sony attack constituted "not just an attack against a company and its employees," but "also an attack on our freedom of expression and way of life." A Seth Rogen comedy may have been an unlikely emblem of the First Amendment and American values; but so were many other works that had come under attack through the nation's history, yet were still worth defending, because an attack on basic values had to be answered—however ignoble the target—lest some future assailant threaten to raid the files of some other studio, publisher, art museum, or record company if their executives didn't cancel some other film, book, exhibition, or album.

The confrontation ticked off a debate inside the Obama White House, similar to the debates discussed, but never resolved, under previous presidents: What *was* a "proportional" response to a cyber attack? Did this response have to be delivered in cyberspace? Finally, what role *should* government play in responding to cyber attacks on citizens or private corporations? A bank gets hacked, that's the bank's problem; but what if two, three, or a dozen banks—big banks—were hacked? At what point did these assaults become a concern for national security?

It was a broader version of the question that Robert Gates had asked the Pentagon's general counsel eight years earlier: at what point did a cyber attack constitute an act of war? Gates never received a clear reply, and the fog hadn't lifted since.

On December 22, three days after Obama talked about the Sony hack at his press conference, someone disconnected North Korea from the Internet. Kim Jong-un's spokesmen accused Washington of launching the attack. It was a reasonable guess: Obama had pledged to launch a "proportional" response to the attack on Sony; shutting down North Korea's Internet for ten hours seemed to fit the bill, and it wouldn't have been an onerous task, given that the whole country had just 1,024 Internet Protocol addresses (fewer than the number on some *blocks* in New York City), all of them connected through a single service provider in China.

In fact, though, the United States government played no part in the shutdown. A debate broke out in the White House over whether to deny the charge publicly. Some argued that it might be good to clarify what a proportional response was *not*. Others argued that making any statement would set an awkward precedent: if U.S. officials issued a denial now, then they'd also have to issue a denial the next time a digital calamity occurred during a confrontation; otherwise everyone would infer that America did launch *that* attack, whether or not it actually had, at which point the victim might fire back.*

In this instance, the North Koreans didn't escalate the conflict, in part because they *couldn't*. But another power, with a more robust Internet, might have.

Gates's question was more pertinent than ever, but it was also, in a sense, beside the point. Because of its lightning speed and the initial ambiguity of its source, a cyber attack could provoke a counterattack, which might escalate to war, in cyberspace and in real space, regardless of anyone's intentions.

At the end of Bush's presidency and the beginning of Obama's, in casual conversations with aides and colleagues in the Pentagon and the White House, Gates took to mulling over larger questions about cyber espionage and cyber war.

"We're wandering in dark territory," he would say on these occasions.

It was a phrase from Gates's childhood in Kansas, where his grandfather worked for nearly fifty years as a stationmaster on the Santa Fe Railroad. "Dark territory" was the industry's term for a stretch of rail track that was uncontrolled by signals. To Gates, it was a perfect parallel to cyberspace, except that this new territory was much vaster and the danger was greater, because the engineers were unknown, the trains were invisible, and a crash could cause far more damage.

Even during the darkest days of the Cold War, Gates would tell his colleagues, the United States and the Soviet Union set and followed

---

*As a compromise, when Obama issued an executive order imposing new sanctions against North Korea, on January 2, 2015, White House spokesman Josh Earnest pointedly called it "the first aspect of our response" to the Sony hacking. Listeners could infer from the word "first" that the United States had not shut down North Korea's Internet eleven days earlier. But no official spelled this out explicitly, at least not on the record.

some basic rules: for instance, they agreed not to kill each other's spies. But today, in cyberspace, there were no such rules, no rules of any kind. Gates suggested convening a closed-door meeting with the other major cyber powers—the Russians, Chinese, British, Israelis, and French—to work out some principles, some "rules of the road," that might diffuse our mutual vulnerabilities: an agreement, say, not to launch cyber attacks on computer networks controlling dams, water-works, electrical power grids, and air traffic control—critical civilian infrastructure—except perhaps in wartime, and maybe not even then.

Those who heard Gates's pitch would furrow their brows and nod gravely, but no one followed up; the idea went nowhere.

Over the next few years, this dark territory's boundaries widened, and the volume of traffic swelled.

In 2014, there were almost eighty thousand security breaches in the United States, more than two thousand of which resulted in losses of data—a quarter more breaches, and 55 percent more data losses, than the year before. On average, the hackers stayed inside the networks they'd breached for 205 days—nearly seven months—before being detected.

These numbers were likely to soar, with the rise of the Inter-net of Things. Back in 1996, Matt Devost, the computer scientist who simulated cyber attacks in NATO war games, co-wrote a paper called "Information Terrorism: Can You Trust Your Toaster?" The title was a bit facetious, but twenty years later, with the most mun-dane items of everyday life—toasters, refrigerators, thermostats, and cars—sprouting portals and modems for network connectivity (and thus for hackers too), it seemed prescient.*

---

*In 2013, two security researchers—including Charlie Miller, a former employee at the Office of Tailored Access Operations, the NSA's elite hacking unit—hacked into the computer system of a Toyota Prius and a Ford Escape, then disabled the brakes and commandeered the steering wheel while the cars were driven around a parking lot. In that test, they'd wired their laptops to the cars' onboard diagnostic ports, which service centers could access online. Two years later, they took control

President Obama tried to stem the deluge. On February 12, 2015, he signed an executive order titled "Improving Critical Infrastructure Cybersecurity," setting up forums in which private companies could share data about the hackers in their midst—with one another and with government agencies. In exchange, the agencies—mainly the NSA, working through the FBI—would provide top secret tools and techniques to protect their networks from future assaults.

These forums were beefed-up versions of the Information Sharing and Analysis Centers that Richard Clarke had established during the Clinton administration—and they were afflicted with the same weakness: both were voluntary; no company executives had to share information if they didn't want to. Obama made the point explicitly: "Nothing in this order," his document stated, "shall be construed to provide an agency with authority for regulating the security of critical infrastructure."

Regulation—it was still private industry's deepest fear, deeper than the fear of losing millions of dollars at the hands of cyber criminals or spies. As the white-hat hacker Peiter "Mudge" Zatko had explained to Dick Clarke fifteen years earlier, these executives had calculated that it cost no more to clean up after a cyber attack than to prevent one in the first place—and the preventive measures might not work anyway.

Some industries had altered their calculations in the intervening

---

of a Jeep Cherokee *wirelessly*, after discovering many vulnerabilities in its onboard computers—which they also hacked wirelessly, through the Internet, cellular channels, and satellite data-links—while a writer for *Wired* magazine drove the car down a highway. Fiat Chrysler, the Jeep's manufacturer, recalled 1.4 million vehicles, but Miller made clear that most, maybe all, modern cars were probably vulnerable in similar ways (though none of them were recalled). As with most other devices in life, their most basic functions had been computerized—and the computers hooked up to networks—for the sake of convenience, their manufacturers oblivious to the dangers they were opening up. The signs of a new dimension in the cyber arms race—involving sabotage, mayhem, terrorism, even assassination plots, carried out more invisibly than drone strikes—seemed ominous and almost inevitable.

years, notably the financial sector. Its business consisted of bringing in money and cultivating trust; hackers had made an enormous dent on both, and sharing information demonstrably lowered risk. But the big banks were exceptions to the pattern.

Obama's cyber policy aides had made a run, early on, at drafting mandatory security standards, but they soon pulled back. Corporate resistance was too stiff; the secretaries of treasury and commerce argued that onerous regulations would impede an economic recovery, the number-one concern to a president digging the country out of its deepest recession in seventy years. Besides, the executives had a point: companies that *had* adopted tight security standards were still getting hacked. The government had offered tools, techniques, and a list of "best practices," but "best" didn't mean perfect—after the hacker adapted, erstwhile best practices might not even be good—and, in any case, tools were just tools: they weren't solutions.

Two years earlier, in January 2013, a Defense Science Board task force had released a 138-page report on "the advanced cyber threat." The product of an eighteen-month study, based on more than fifty briefings from government agencies, military commands, and private companies, the report concluded that there was no reliable defense against a resourceful, dedicated cyber attacker.

In several recent exercises and war games that the panel reviewed, Red Teams, using exploits that any skilled hacker could download from the Internet, "invariably" penetrated even the Defense Department's networks, "disrupting or completely beating" the Blue Team.

The outcomes were all too reminiscent of Eligible Receiver, the 1997 NSA Red Team assault that first exposed the U.S. military's abject vulnerability.

Some of the task force members had observed up close the early history of these threats, among them Bill Studeman, the NSA director in the late 1980s and early 1990s, who first warned that the agency's radio dishes and antennas were "going deaf" in the global transition from analog to digital; Bob Gourley, one of Studeman's acolytes, the first intelligence chief of the Pentagon's Joint Task Force-Computer Network Defense, who traced the Moonlight Maze hack to Russia; and Richard Schaeffer, the former director of the NSA Information Assurance Directorate, who spotted the first known penetration of the U.S. military's *classified* network, prompting Operation Buckshot Yankee.

Sitting through the briefings, collating their conclusions, and writing the report, these veterans of cyber wars past—real and simulated—felt as if they'd stepped into a time machine: the issues, the dangers, and, most surprising, the vulnerabilities were the same as they'd been all those years ago. The government had built new systems and software, and created new agencies and directorates, to detect and resist cyber attacks; but as with any other arms race, the offense—at home and abroad—had devised new tools and techniques as well, and, in this race, the offense held the advantage.

"The network connectivity that the United States has used to tremendous advantage, economically and militarily, over the past twenty years," the report observed, "has made the country more vulnerable than ever to cyber attacks." It was the same paradox that countless earlier commissions had observed.

The problem was basic and inescapable: the computer networks, the panelists wrote, were "built on inherently insecure architectures." The key word here was *inherently*.

It was the problem that Willis Ware had flagged nearly a half century earlier, in 1967, just before the rollout of the ARPANET: the very existence of a computer *network*—where multiple users could gain access to files and data online, from remote, unsecured locations—created inherent vulnerabilities.

The danger, as the 2013 task force saw it, wasn't that someone would launch a cyber attack, out of the blue, on America's military machine or critical infrastructure. Rather, it was that cyber attacks would be an element of all future conflicts; and given the U.S. military's dependence on computers—in everything from the GPS guidance systems in its missiles, to the communications systems in its command posts, to the power stations that generated its electricity, to the scheduling orders for resupplying the troops with ammunition, fuel, food, and water—there was no assurance that America would win this war. "With present capabilities and technology," the report stated, "it is not possible to defend with confidence against the most sophisticated cyber attacks."

Great Wall defenses could be leapt over or maneuvered around. Instead, the report concluded, cyber security teams, civilian and military, should focus on *detection* and *resilience*—designing systems that could spot an attack early on and repair the damage swiftly.

More useful still would be figuring out ways to *deter* adversaries from attacking even in the most tempting situations.

This had been the great puzzle in the early days of nuclear weapons, when strategists realized that the atomic bomb and, later, the hydrogen bomb were more destructive than any war aim could justify. As Bernard Brodie, the first nuclear strategist, put it in a book called *The Absolute Weapon*, published just months after Hiroshima and Nagasaki, "Thus far the chief purpose of our military establishment has been to win wars. From now on its chief purpose must be to avert them." The way to do that, Brodie reasoned, was to protect the nuclear arsenal, so that, in the event of a Soviet first strike, the United States would have enough bombs surviving to "retaliate in kind."

But what did that mean in modern cyberspace? The nations most widely seen as likely foes in such a war—Russia, China, North Korea, Iran—weren't plugged into the Internet to nearly the same extent as America. Retaliation in kind would inflict far less damage on those countries than the first strike had inflicted on America; therefore, the *prospect* of retaliation might not deter them from attacking. So what *was* the formula for cyber deterrence: threatening to respond to an attack by declaring all-out war, firing missiles and smart bombs, escalating to nuclear retaliation? Then what?

The fact was, no one in a position of power or high-level influence had thought this through.

Mike McConnell had pondered the question in the transition between the Bush and Obama presidencies, when he set up the Comprehensive National Cybersecurity Initiative. The CNCI set twelve tasks to accomplish in the ensuing few years: among other things, to install a common intrusion-detection system across all federal networks, boost the security of classified networks, define the U.S. government's role in protecting critical infrastructure—and there was this (No. 10 on the list): "Define and develop enduring deterrence strategies and programs."

Teams of aides and analysts were formed to work on the twelve projects. The team assigned to Task No. 10 came up short: a paper was written, but its ideas were too vague and abstract to be described as "strategies," much less "programs."

McConnell realized that the problem was too hard. The other tasks were hard, too, but in most of those cases, it was fairly clear *how* to get the job done; the trick was getting the crucial parties—the bureaucracies, Congress, and private industry—to do it. Figuring out cyber deterrence was a conceptual problem: which hackers were you trying to deter; what were you trying to deter them from doing; what penalties were you threatening to impose if they attacked anyway; and how would you make sure they wouldn't strike back harder in response? These were questions for policymakers, maybe political philosophers, not for midlevel aides on a task force.

The 2013 Defense Science Board report touched lightly on the question of cyber deterrence, citing parallels with the advent of the A-bomb at the end of World War II. "It took decades," the report noted, "to develop an understanding" of "the strategies to achieve stability with the Soviet Union." Much of this understanding grew out of analyses and war-game exercises at the RAND Corporation, the Air Force–sponsored think tank where civilian economists, physicists, and political scientists—among them Bernard Brodie—conceived and tested new ideas. "Unfortunately," the task force authors wrote, they "could find no evidence" that anyone, anywhere, was doing that sort of work "to better understand the large-scale cyber war."

The first official effort to find some answers to these questions got underway two years later, on February 10, 2015, with the opening session of yet another Defense Science Board panel, this one called the Task Force on Cyber Deterrence. It would continue meeting in a highly secure chamber in the Pentagon for two days each month, through the end of the year. Its goal, according to the memo that created the panel, was "to consider the requirements for effective deterrence of cyber attack against the United States and allies/partners."

Its panelists included a familiar group of cyber veterans, among them Chris Inglis, deputy director of the NSA under Keith Alexander, now a professor of cyber studies at the U.S. Naval Academy in Annapolis, Maryland; Art Money, the former Pentagon official who guided U.S. policy on information warfare in the formative era of the late 1990s, now (and for the previous decade) chairman of the NSA advisory board; Melissa Hathaway, the former Booz Allen project manager who was brought into the Bush White House by Mike McConnell to run the Comprehensive National Cybersecurity Initiative, now the head of her own consulting firm; and Robert Butler, a former officer at the Air Force Information Warfare Center who'd helped run the first modern stab at information warfare, the campaign against Serbian president Slobodan Milosevic and his cronies. The chairman of the task force was James Miller, the undersecretary of defense for policy, who'd been working cyber issues in the Pentagon for more than fifteen years.

All of them were longtime inside players of an insiders-only game; and, judging from their presence, the Pentagon's permanent bureaucrats wanted to keep it that sort of game.

Meanwhile, the power and resources were concentrated at Fort Meade, where U.S. Cyber Command was amassing its regiments, and drawing up battle plans, even though broad questions of policy and guidance had barely been posed, much less settled.

In 2011, when Robert Gates realized that the Department of Homeland Security would never be able to protect the nation's critical infrastructure from a cyber attack (and after his plan for a partnership between DHS and the NSA went up in smoke), he gave that responsibility to Cyber Command as well.

Cyber Command's original two core missions were more straightforward. The first, to support U.S. combatant commanders, meant going through their war plans and figuring out which targets could be destroyed by cyber means rather than by missiles, bullets, or bombs. The second mission, to protect Defense Department computer networks, was right up Fort Meade's alley: those networks had only eight points of access to the Internet; Cyber Command could sit on all of them, watching for intruders; and, of course, it had the political and legal authority to monitor, and roam inside, those networks, too.

But its third, new mission—defending civilian critical infrastructure—was another matter. The nation's financial institutions, power grids, transportation systems, waterworks, and so forth had thousands of access points to the Internet—no one knew precisely how many. And even if the NSA could somehow sit on those points, it lacked the legal authority to do so. Hence Obama's executive order, which relied on private industry to share information voluntarily— an unlikely prospect, but the only one available.

It was a bitter irony. The growth of this entire field—cyber security, cyber espionage, cyber war—had been triggered by concerns, thirty years earlier, about the vulnerability of critical infrastructure. Yet, after all the commissions, analyses, and directives, the problem seemed intractable.

Still, Keith Alexander not only accepted the new mission, he aggressively pushed for it; he'd helped Gates draft the directive that gave the mission to Cyber Command. To Alexander's mind, not only did Homeland Security lack the resources to protect the nation, it had the wrong concept. It was trying to install intrusion-detection systems on all the networks, and there were just too many networks: they'd be impossible to monitor, and it would cost way too much to try. Besides, what could the DHS bureaucrats do if they detected a serious attack in motion?

The better approach, to Alexander's mind, was the one he knew best: to go on the offensive—to get inside the adversary's networks in order to see him preparing an attack, then deflect it. This was the age-old concept of "active defense" or, in its cyber incarnation, CNE, Computer Network Exploitation, which, as NSA directors dating back to Ken Minihan and Mike Hayden knew well, was not much different from Computer Network Attack.

But Alexander advocated another course, too, a necessary supplement: *force* the banks and the other sectors—or ply them with alluring incentives—to share information about their hackers with the government: and by "government," he meant the FBI and, through it, the NSA and Cyber Command. He decidedly did not mean the Department of Homeland Security—though, in deference to the White House, which had designated DHS as the lead agency on protecting critical infrastructure, he would say the department could act as the "router" that sent alerts to the other, more active agencies.

Alexander was insistent on this point. Most private companies refused to share information, not only because they lacked incentives but also because they feared lawsuits: some of that information would include personal data about employees and customers. In response, President Obama urged Congress to pass a bill exempting companies from liability if they shared data. But Alexander opposed the bill, because Obama's version of the bill would require them to share data with the Department of Homeland Security. Without telling the White House, Alexander lobbied his allies on Capitol Hill to amend or kill his commander-in-chief's initiative.

It was an impolitic move from someone who was usually a bit more adroit. First, the White House staff soon heard about his lobbying, which didn't endear him to the president, especially in the wake of the Snowden leaks, which were already cutting into the reserves of goodwill for Fort Meade. Second, it was self-defeating from a substantive angle: even with exemption from liability, companies were averse to giving private data to the government—all the more so if "government" was openly defined as the NSA.

The information-sharing bill was endangered, then, by an unlikely coalition of civil liberties advocates, who opposed sharing data with the government on principle, and NSA boosters, who opposed sharing it with any entity but Fort Meade.

So, the only coordinated defense left would be "active defense"—cyber offensive warfare.

That was the situation inherited by Admiral Michael Rogers, who replaced Alexander in April 2014. A career cryptologist, Rogers had run the Navy's Fleet Cyber Command, which was also based at Fort Meade, before taking over the NSA and U.S. Cyber Command. He was also the first naval officer to earn three stars (and now he had four stars) after rising through the ranks as a code-breaker. Shortly after taking the helm, he was asked, in an interview with the Pentagon's news service, how he would protect critical infrastructure from a cyber attack—Cyber Command's third mission. He replied that the "biggest focus" would be "to attempt to interdict the attack before it ever got to us"—in other words, to get inside the adversary's network, in order to see him prepare an attack, then to deflect or preempt it.

"Failing that," Rogers went on, he would "probably" also "work directly with those critical infrastructure networks" that "could use stronger defensive capabilities." But he knew this was backup, and flimsy backup at that, since neither Fort Meade nor the Pentagon could do much to bolster the private sector's defenses on its own.

In April 2015, the Obama administration endorsed the logic. In a thirty-three-page document titled *The Department of Defense Cyber Strategy*, signed by Ashton Carter, a former Harvard physicist, longtime Pentagon official, and now Obama's fourth secretary of defense, the same three missions were laid out in some detail: assisting the U.S. combatant commands, protecting Defense Department networks, and protecting critical infrastructure. To carry out this last mission, the document stated that, "with other government agencies" (the standard euphemism for NSA), the Defense Department had developed "a range of options and methods for disrupting cyber attacks of significant consequence *before* they can have an impact." And it added, in a passage more explicit than the usual allusions to the option of Computer Network Attack, "If directed, DoD should be able to use cyber operations to disrupt an adversary's command-and-control networks, military-related critical infrastructure, and weapons capabilities."

A month earlier, on March 19, at hearings before the Senate Armed Services Committee, Admiral Rogers expressed the point more directly still, saying that deterring a cyber attack required addressing the question: "How do we increase our capacity on the offensive side?"

Senator John McCain, the committee's Republican chairman, asked if it was true that the "current level of deterrence is not deterring."

Rogers replied, "That is true." More cyber deterrence meant more cyber offensive tools and more officers trained to use them, which meant more money and power for Cyber Command.

But *was* this true? At an earlier hearing, Rogers had made headlines by testifying that China and "probably one or two other countries" were definitely inside the networks that controlled America's power grids, waterworks, and other critical assets. He didn't say so, but America was also inside the networks that controlled such assets

in those other countries. Would burrowing more deeply deter an attack, or would it only tempt both sides, all sides, to attack the others'
networks preemptively, in the event of a crisis, before the other sides
attacked their networks first? And once the exchanges got under way,
how would anyone keep them from escalating to more damaging
cyber strikes or to all-out war?

These were questions that some tried to answer, but no one ever
did, during the nuclear debates and gambits of the Cold War. But
while nuclear weapons were incomparably more destructive, there
were four differences about this new arms race that made it more
likely to career out of control. First, more than two players were
involved, a few were unpredictable, and some weren't even nation-states. Second, an attack would be invisible and, at first, hard
to trace, boosting the chances of mistakes and miscalculations on
the part of the country first hit. Third, a bright, bold firewall separated using nuclear weapons from not using nuclear weapons; the
countries that possessed the weapons were constrained from using
them, in part, because no one knew how fast and furious the violence
would spiral, once the wall came down. By contrast, cyber attacks of
one sort or another were commonplace: they erupted more than two
hundred times a day, and no one knew—no one had ever declared,
no one could predict—where the line between mere nuisance and
grave threat might be drawn; and so there was a higher chance that
someone would cross the line, perhaps without intending or even
knowing it.

Finally, there was the extreme secrecy that enveloped everything
about cyber war. Some things about nuclear weapons were secret,
too: details about their design, the launch codes, the targeting plans,
the total stockpile of nuclear materials. But the basics were well
known: their history, how they worked, how many there were, how
much destruction they could wreak—enough to facilitate an intel-
ligent conversation, even by people who didn't have Top Secret se-

curity clearances. This was not true of cyber: when Admiral Rogers testified that he wanted to "increase our capacity on the offensive side," few, if any, of the senators had the slightest idea what he was talking about.

In the five guys report on NSA reform, which President Obama commissioned in 2013 in the wake of the Snowden revelations, the authors acknowledged, even stressed, the need to keep certain sources, methods, and operations highly classified. But they also approvingly quoted a passage from the report by Senator Frank Church, written in the wake of another intelligence scandal—that one, clearly illegal—almost forty years earlier. "The American public," he declared, "should know enough about intelligence activities to be able to apply their good sense to the underlying issues of policy and morality."

This knowledge, which Senator Church called "the key to control," has been missing from discussions of policy, strategy, and morality in cyber war. We are all wandering in dark territory, most of us only recently, and even now dimly, aware of it.

NOTES

MUCH OF the material in this book comes from interviews, all conducted on background, with more than a hundred participants in the story, many of them followed up with email, phone calls, or repeated in-person interviews. (For more about these sources, see the Acknowledgments.) In the Notes that follow, I have not cited sources for material that comes strictly from interviews. For material that comes in part from written sources (books, articles, documents, and so forth) and in part from interviews, I have cited those sources, followed by "and interviews."

CHAPTER 1: "COULD SOMETHING LIKE THIS REALLY HAPPEN?"

1 *That night's feature:* During his eight years as president, at Camp David and in the White House screening room, Reagan watched 374 movies, an average of nearly one a week, though often more. ("Movies Watched at Camp David and White House," Aug. 19, 1988, 1st Lady Staff Office Papers, Ronald Reagan Library.) *WarGames* was an unusual choice; he usually watched adventures, light comedies, or musicals. But one of the film's screenwriters, Lawrence Lasker, was the son of the actress Jane Greer and the producer Edward Lasker, old friends of Reagan from his days as a Hollywood movie star. Lawrence used his family connections to get a print to the president. (Interviews.)

1 *The following Wednesday morning:* Office of the President, Presidential Briefing Papers, Box 31, 06/08/1983 (case file 150708) (1), Ronald Reagan Library; and interviews. This meeting is mentioned in Lou Cannon, *President Reagan: The Role of a Lifetime* (New York: Simon & Schuster, 1991), 38, but, in addition to getting the date wrong, Cannon depicts it as just another wacky case of Reagan taking movies too seriously; he doesn't recount the

president's question to Gen. Vessey, nor does he seem aware that the viewing and this subsequent White House meeting had an impact on history. See also Michael Warner, "Cybersecurity: A Pre-history," *Intelligence and National Security*, Oct. 2012.

2 *"highly susceptible to interception":* NSDD-145 has since been declassified: http://fas.org/irp/offdocs/nsdd145.htm.

3 *Established in 1952:* As later codified in Executive Order 12333, signed by Ronald Reagan on Dec. 4, 1981, the NSA and FBI were barred from undertaking foreign intelligence collection "for the purpose of acquiring information concerning the domestic activities of United States persons," this last phrase referring to American citizens, legal residents, and corporations (http://www.archives.gov/federal-register/codification/executive-or der/12333.html).

4 *In its first three years:* Ellen Nakashima, "Pentagon to Boost Cybersecurity Force," *Washington Post*, Jan. 27, 2013; and interviews.

4 *In the American Civil War:* Edward J. Glantz, "Guide to Civil War Intelligence," *The Intelligencer: Journal of U.S. Intelligence Studies* (Winter/Spring 2011), 57; Jason Healey, ed., *A Fierce Domain: Conflict in Cyberspace, 1986 to 2012* (Washington, D.C.: Atlantic Council, 2013), 27.

4 *During World War II:* See esp. David Kahn, *The Codebreakers* (New York: Scribner; rev. ed., 1996), Ch. 14.

6 *a man named Donald Latham:* Warner, "Cybersecurity: A Pre-history"; and interviews.

8 *In April 1967:* Willis H. Ware, *Security and Privacy in Computer Systems* (Santa Monica: RAND Corporation, P-3544, 1967). This led to a 1970 report by a Defense Science Board task force, known as "the Ware Panel," *Security Controls for Computer Systems* (declassified by RAND Corporation as R-609-1, 1979); and interviews.

8 *He well understood:* Willis H. Ware, *RAND and the Information Evolution: A History in Essays and Vignettes* (Santa Monica: RAND Corporation, 2008).

8 *Ware was particularly concerned:* Ibid., 152ff.

9 *In 1980, Lawrence Lasker and Walter Parkes:* Extra features, *WarGames: The 25th Anniversary Edition*, Blu-ray disc; and interviews.

11 *The National Security Agency had its roots:* See Kahn, *The Codebreakers*, 352. The stories about the tenth floor of the embassy and Inman's response to reports of a fire are from interviews. The fact that U.S. intelligence was listening in on Brezhnev's limo conversations (though not its method) was revealed by Jack Anderson, "CIA Eavesdrops on Kremlin Chiefs," *Washington Post*, Sept. 16, 1971. Anderson's source was a right-wing Senate aide who argued

that the transcripts proved the Russians were cheating on the latest nuclear arms control treaty. After Anderson's story appeared, the Russians started encrypting their phone conversations. The NSA broke the codes. Then the Russians installed more advanced encryption, and that was the end of the operation. (All this backstory is from interviews.)

16 *In his second term as president:* Don Oberdorfer, *From the Cold War to a New Era* (Baltimore: Johns Hopkins University Press, 1998), 67.

16 *When they found out about the microwaves:* Associated Press, "Russia Admits Microwaves Shot at US Embassy," July 26, 1976; "Science: Moscow Microwaves," *Time*, Feb. 23, 1976. The news stories note that personnel on the tenth floor were experiencing health problems due to the microwave beams. The stories don't reveal—probably the reporters didn't know—the purpose of the beams (they quote embassy officials saying they're baffled about them) or the activities on the tenth floor.

16 *took to playing Muzak:* As a defense reporter for *The Boston Globe* in the 1980s, I often heard Muzak when I interviewed senior Pentagon officials in their offices. I asked one of them why it was playing. He pointed to his window, which overlooked the Potomac, and said the Russians might be listening with microwave beams.

CHAPTER 2: "IT'S ALL ABOUT THE INFORMATION"

26 *Its number-one mission:* Most of this is from interviews, but see also Christopher Ford and David Rosenberg, *The Admirals' Advantage: U.S. Navy Operational Intelligence in World War II and the Cold War* (Annapolis: Naval Institute Press, 2005), esp. Ch. 5. (All the material about Desert Storm is from interviews.)

31 *McConnell sat up as he watched:* Though *Sneakers* inspired McConnell to call the concept "information warfare," the phrase had been used before, first by weapons scientist Thomas P. Rona in a Boeing Company monograph, "Weapon Systems and Information War" (Boeing Aerospace Company, July 1976). Rona was referring not to computers but to technology that theoretically enhanced the capability of certain weapons systems by linking them to intelligence sensors.

32 *"decapitate the enemy's command structure":* Warner, "Cybersecurity: A Prehistory."

36 *McConnell pushed hard for the Clipper Chip:* Jeffrey R. Yost, "An Interview with Dorothy E. Denning," OH 424, Computer Security History Project, April 11, 2013, Charles Babbage Institute, University of Minnesota,

http://conservancy.umn.edu/bitstream/handle/11299/156519/oh424ded
.pdf?sequence=1; and interviews.

CHAPTER 3: A CYBER PEARL HARBOR

40 *"critical national infrastructure":* President Bill Clinton, PDD-39, "U.S. Policy on
    Counterterrorism," June 21, 1995, http://fas.org/irp/offdocs/pdd/pdd-39.pdf.
40 *Reno turned the task over:* Most of the material on the Critical Infrastructure
    Working Group comes from interviews with several participants, though
    some is from Kathi Ann Brown, *Critical Path: A Brief History of Critical In-
    frastructure Protection in the United States* (Fairfax, VA: Spectrum Publishing
    Group, 2006), Chs. 5, 6. All details about briefings and private conversations
    within the group come from interviews.
42 *"high-tech matters":* Memo, JoAnn Harris, through Deputy Attorney General
    [Jamie Gorelick] to Attorney General, "Computer Crime Initiative Action
    Plan," May 6, 1994; Memo, Deputy Attorney General [Gorelick], "Forma-
    tion of Information Infrastructure Task Force Coordinating Committee,"
    July 19, 1994 (provided to author); and interviews.
42 *In recent times: Security in Cyberspace: Hearings Before the Permanent Subcommittee
    on Investigations of the Comm. on Government Affairs.* 104th Cong. (1996). (state-
    ment of Jamie Gorelick, Deputy Attorney General of the United States.)
42 *the interagency meetings with Bill Studeman:* Studeman's role on interagency
    panels comes from Douglas F. Garthoff, *Directors of Central Intelligence as
    Leaders of the U.S. Intelligence Community, 1946–2005* (Washington, D.C.: CIA
    Center for the Study of Intelligence, 2005), 267. That he and Gorelick met
    every two weeks was noted in *Security in Cyberspace: Hearings Before the Per-
    manent Subcommittee on Investigations of the Comm. on Government Affairs.* 104th
    Cong. (1996). (statement of Jamie Gorelick, Deputy Attorney General of the
    United States.)
43 *One branch of J Department:* "Critical nodes" theory has fallen short in real-life
    wars. The Air Force attack plan for the 1990–91 Gulf War focused on eighty-
    four targets as the key "nodes": destroy those targets, and the regime would
    collapse like a house of cards. In fact, the war didn't end until a half mil-
    lion U.S. and allied troops crushed Iraq's army on the ground. See Michael
    Gordon and Bernard Trainor, *The Generals' War* (New York: Little, Brown,
    1995), Ch. 4; Fred Kaplan, *Daydream Believers* (Hoboken: John Wiley & Sons,
    2008), 20–21.
45 *Capping Greene's briefing, the CIA:* Brown, *Critical Path*, 78; and interviews.
45 *"in light of the breadth":* This language was reproduced in a memorandum

from Attorney General to the National Security Council, on March 16, http://fas.org/sgp/othergov/munromem.htm.

45 *One word was floating around:* The first use of "cyber war" was probably John Arquilla and David Ronfeldt, *Cyberwar Is Coming!* (Santa Monica: RAND Corporation, 1993), but their use of the phrase was more like what came to be called "netcentric warfare" or the "revolution in military affairs," not "cyber war" as it later came to be understood.

47 *"may have experienced as many as 250,000 attacks":* General Accounting Office, "Information Security: Computer Attacks at Department of Defense Pose Increasing Risks" (GAO/AIMD-96-84), May 22, 1996. The report attributes the estimate to a study by the Pentagon's Defense Information Security Agency.

47 *"Certain national infrastructures":* President Bill Clinton, Executive Order 13010, "Critical Infrastructure Protection," July 15, 1996, http://fas.org/irp/offdocs/eo13010.htm.

48 *"We have not yet had a terrorist":* Jamie Gorelick, *Security in Cyberspace: Hearings Before the Permanent Subcommittee on Investigations of the Comm. on Government Affairs.* 104th Cong. (1996) (Statement of Jamie Gorelick, Deputy Attorney General of the United States.)

48 *America's programs in this realm:* There were only a few slipups in revealing the existence of a cyber offensive program, and they were little noticed. In May 1995, Emmett Paige, assistant secretary of defense for command, control, communications, and intelligence, said at a conference at the National Defense University, "We have an offensive [cyber] capability, but we can't discuss it. . . . You'd feel good about it if you knew about it." The next month, Navy Captain William Gravell, director of the Joint Staff's information warfare group, said at a conference in Arlington, "We are at the first stage of a comprehensive effort [in information warfare]. . . . What we have been doing up to now is building some very powerful offensive systems." As for now, he added, "there is no current policy in these matters." That would remain true for many years after. Both remarks were quoted in Neil Munro, "Pentagon Developing Cyberspace Weapons," *Washington Technology*, June 22, 1995—with no follow-up in any mass media, http://washingtontechnology.com/Articles/1995/06/22/Pentagon-Developing-Cyberspace-Weapons.aspx.

51 *Marsh and the commissioners first convened:* Brown, *Critical Path,* 93. The rest of the material on the commission comes from interviews.

53 *"Just as the terrible long-range weapons":* White House, *Critical Foundations: Protecting America's Infrastructures: The Report of the President's Commission on Critical Infrastructure Protection,* Oct. 1997, http://fas.org/sgp/library/pccip.pdf.

54 *"a serious threat to communications infrastructure":* Commission on Engineering

and Technical Systems, National Research Council, *Growing Vulnerability of the Public Switched Networks: Implications for National Security Emergency Preparedness* (Washington, D.C.: National Academy Press, 1989), 9.

54   *"The modern thief":* Commission on Engineering and Technical Systems, National Research Council, *Computers at Risk: Safe Computing in the Information Age* (Washington, D.C.: National Academy Press, 1991), 7.

54   *"increasing dependency":* *Report of the Defense Science Board Task Force on Information Warfare-Defense* (Washington, D.C.: Office of the Undersecretary of Defense [Acquisition and Technology], 1996). Quotes are from Duane Andrews, cover letter to Craig Fields, Nov. 27, 1996.

55   *"In our efforts to battle":* Transcript, President Bill Clinton, Address to Naval Academy, Annapolis, MD, May 22, 1998, http://www.cnn.com/ALLPOLI TICS/1998/05/22/clinton.academy/transcript.html.

CHAPTER 4: ELIGIBLE RECEIVER

57   *On June 9, 1997:* Most of the material on Eligible Receiver comes from interviews with participants, but some also comes from these printed sources: Brig. Gen. Bruce Wright, "Eligible Receiver 97," PowerPoint briefing, n.d. (declassified; obtained from the Cyber Conflict Studies Association); Dillon Zhou, "Findings on Past US Cyber Exercises for 'Cyber Exercises: Yesterday, Today and Tomorrow'" (Washington, D.C.: Cyber Conflict Studies Association, March 2012); Warner, "Cybersecurity: A Pre-history."

60   *The first nightmare case:* For more on the Morris Worm, see Cliff Stoll, *The Cuckoo's Egg* (New York: Doubleday, 1989), 385ff; Mark W. Eichin and Jon A. Rochlis, "With Microscope and Tweezers: An Analysis of the Internet Virus of November 1988" (MIT, Feb. 9, 1989), presented at the 1989 IEEE Symposium on Research in Security and Privacy, http://www.utdallas. edu/~edsha/UGsecurity/internet-worm-MIT.pdf.

60   *Todd Heberlein's innovation:* Richard Bejtlich, *The Practice of Network Security Monitoring* (San Francisco: No Starch Press, 2013), esp. the foreword (by Todd Heberlein) and Ch. 1; Richard Bejtlich, TAO Security blog, "Network Security Monitoring History," April 11, 2007, http://taosecurity.blogspot .com/2007/04/network-security-monitoring-history.html; and interviews. Bejtlich, who was an officer at the Air Force Information Warfare Center, later became chief security officer at Mandiant, one of the leading private cyber security firms. The founding president, Kevin Mandia, rose through Air Force ranks as a cyber crime specialist at the Office of Special Investigations;

during that time, he frequently visited AFIWC, where he learned of—and was greatly influenced by—its network security monitoring system.

61 *A junior officer:* That was Bejtlich. See a version of his review at http://www .amazon.com/review/RLLSEQRTT5DIF.

63 *"banner warning":* Letter, Robert S. Mueller III, Assistant Attorney General, Criminal Division, to James H. Burrows, Director, Computer Systems Laboratory, National Institute of Standards and Technology, Department of Commerce, Oct. 7, 1992, http://www.netsq.com/Documents_html /DOJ_1992_letter/.

64 *by the time he left the Pentagon:* Bejtlich, "Network Security Monitoring History."

66 *These systems had to clear a high bar:* In the 1980s, the Information Assurance Directorate's Computer Security Center wrote a series of manuals, setting the standards for "trusted computer systems." The manuals were called the "Rainbow Series," for the bright colors of their covers. The key book was the first one, the so-called Orange Book, "Trusted Computer Systems Evaluation Criteria," published in 1983. Most of the work was done by the Center's director, Roger Schell, who, a decade earlier, had helped the intelligence community penetrate adversary communications systems and thus knew that U.S. systems would soon be vulnerable too.

67 *On February 16, 1997:* CJCS Instruction No. 3510.01, "No-Notice Interoperability Exercise (NIEX) Program," quoted in Zhou, "Findings on Past US Cyber Exercises for 'Cyber Exercises: Yesterday, Today and Tomorrow.'"

67 *The game laid out a three-phase scenario:* Wright, "Eligible Receiver 97," PowerPoint briefing, The rest of the section is based on interviews with participants.

69 *The person answering the phone:* Matt Devost of the Coalition Vulnerability Assessment Team had experienced similar problems when he tried to find the American commander's computer password during one of the five eyes nations' war games. First, he unleashed a widely available software program that, in roughly one second's time, tried out every word in the dictionary with variations. Then he phoned the commander's office, said he was with a group that wanted him to come speak, and asked for a biographical summary. He used the information on that sheet to generate new passwords, and broke through with "Rutgers" (where the commander's son was going to college) followed by a two-digit number.

72 *it only briefly alluded to:* White House, *Critical Foundations: Protecting America's Infrastructures: The Report of the President's Commission on Critical Infrastructure Protection,* Oct. 1997, 8, http://fas.org/irp/offdocs/nsdd145.htm.

CHAPTER 5: SOLAR SUNRISE, MOONLIGHT MAZE

73　*On February 3, 1998:* The tale of Solar Sunrise comes mainly from interviews but also from Richard Power, "Joy Riders: Mischief That Leads to Mayhem," *InforMIT*, Oct. 30, 2000, http://www.informit.com/articles/article.as px?p=19603&seqNum=4; *Solar Sunrise: Dawn of a New Threat*, FBI training video, www.wired.com/20087/09/video-solar-sun/; Michael Warner, "Cybersecurity: A Pre-history;" and sources cited below.

74　*"the first shots":* Bradley Graham, "US Studies a New Threat: Cyber Attack," *Washington Post*, May 24, 1998.

75　*"concern that the intrusions":* FBI, Memo, NID/CID to all field agents, Feb. 9, 1998 (declassified, obtained from the Cyber Conflict Studies Association).

77　*"going to retire":* Power, "Joy Riders."

78　*"the most organized":* Rajiv Chandrasekaran and Elizabeth Corcoran, "Teens Suspected of Breaking into U.S. Computers," *Washington Post*, Feb. 28, 1998.

78　*Israeli police arrested Tenenbaum:* Dan Reed and David L. Wilson, "Whiz-Kid Hacker Caught," *San Jose Mercury News*, March 19, 1998, http://web.archive .org/web/20001007150311/http://www.mercurycenter.com/archives/reprints/hacker110698.htm; Ofri Ilany, "Israeli Hacker Said Behind Global Ring That Stole Millions," *Haaretz*, Oct. 6, 2008, http://www.haaretz.com/print-edition/news/israeli-hacker-said-behind-global-ring-that-stole-mil lions-1.255053.

78　*"not more than the typical hack":* FBI, Memo, [sender and recipient redacted], "Multiple Intrusions at DoD Facilities," Feb. 12, 1998 (obtained from the Cyber Conflict Studies Association files).

81　*"Who's in charge?":* "Lessons from Our Cyber Past—The First Military Cyber Units," symposium transcript, Atlantic Council, March 5, 2012, http://www. atlanticcouncil.org/news/transcripts/transcript-lessons-from-our-cyber-past-the-first-military-cyber-units.

82　*"responsible for coordinating":* Maj. Gen. John H. Campbell, PowerPoint presentation, United States Attorneys' National Conference, June 21, 2000.

82　*Meanwhile, the FBI was probing all leads:* See the many FBI memos, to and from various field offices, in the declassified documents obtained by the Cyber Conflict Studies Association.

86　*5.5 gigabytes of data:* The figure of 5.5 gigabytes comes from Maj. Gen. John H. Campbell, PowerPoint briefing on computer network defense, United States Attorneys' National Conference, June 21, 2000.

86　*Days later, the news leaked to the press:* "Cyber War Underway on Pentagon

Computers—Major Attack Through Russia," CNN, March 5, 1999; Bar-
bara Starr, "Pentagon Cyber-War Attack Mounted Through Russia," ABC
News, March 5, 1999, http://www.rense.com/politics2/cyberwar.htm.

87 *They flew to Moscow on April 2:* Declassified FBI memos, in the files of the
Cyber Conflict Studies Association, mention the trip: for instance, FBI,
Memo, from NatSec, "Moonlight Maze," March 31, 1999; FBI, Memo
(names redacted), Secret/NoForn, "Moonlight Maze Coordinating Group,"
April 15, 1999. The rest of the material comes from interviews. (The April
15 memo also mentions that Justice and Defense Department officials, including Michael Vatis and Soup Campbell, briefed key members of House
and Senate Intelligence Committees on Feb. 21, 1999, and that the first public mention of Moonlight Maze was made by John Hamre on March 5, 1999,
one year after the first intrusions.)

CHAPTER 6: THE COORDINATOR MEETS MUDGE

92 *The collective had started:* The section on Mudge and the L0pht comes mainly
from interviews, though also from Bruce Gottlieb, "HacK, CouNterHaCk,"
New York Times, Oct. 3, 1999; Michael Fitzgerald, "L0pht in Transition,"
CSO, April 17, 2007, http://www.csoonline.com/article/2121870/net
work-security/lopht-in-transition.html; "Legacy of the L0pht," *IT Security Guru*, http://itsecurityguru.org/gurus/legacy-l0pht/#.VGE-CIvF_QU.
Clarke later wrote a novel, *Breakpoint* (New York: G. P. Putnam's Sons,
2007), in which one of the main characters, "Soxster," is based on Mudge;
and a hacker underground called "the Dugout" is modeled on the L0pht.

94 *He'd been a hacker:* His guitar playing at Berklee comes from Mark Small,
"Other Paths: Some High-Achieving Alumni Have Chosen Career Paths
That Have Led Them to Surprising Places," *Berklee*, Fall 2007, http://www.
berklee.edu/bt/192/other_paths.html.

95 *He and the other L0pht denizens:* The hearing can be seen on YouTube, http://
www.youtube.com/watch?v=VVJldn_MmMY.

95 *Three days after Mudge's testimony:* Bill Clinton, Presidential Decision Direc-
tive/NSC-63, "Critical Infrastructure Protection," May 22, 1998, http://fas
.org/irp/offdocs/pdd/pdd-63.htm.

100 *FIDNET, as he called it:* John Markoff, "U.S. Drawing Plan That Will Monitor Computer Systems," *New York Times*, July 28, 1999; and interviews.

101 *"Orwellian":* Tim Weiner, "Author of Computer Surveillance Plan Tries to
Ease Fears," *New York Times*, Aug. 16, 1999; and interviews.

101 *"While the President and Congress can order":* Bill Clinton, *National Plan for*

Information Systems Protection, Jan. 7, 2000, http://cryptome.org/cybersec
-plan.htm.

102 *Still, Clarke persuaded the president to hold a summit:* Most of this comes
from interviews, but see also Gene Spafford, "Infosecurity Summit at the
White House," Feb. 2000, http://spaf.cerias.purdue.edu/usgov/pres.html;
CNN, *Morning News*, Feb. 15, 2000, http://transcripts.cnn.com/TRAN-
SCRIPTS/0002/15/mn.10.html; Ricardo Alonso-Zaldivar and Eric Licht-
blau, "High-Tech Industry Plans to Unite Against Hackers," *Los Angeles
Times*, Feb. 16, 2000.

103 *A few weeks earlier, Mudge had gone legit:* Kevin Ferguson, "A Short, Strange
Trip from Hackers to Entrepreneurs," *Businessweek Online Frontier*, March
2, 2000, http://www.businessweek.com/smallbiz/0003/ep000302.htm?script
framed.

CHAPTER 7: DENY, EXPLOIT, CORRUPT, DESTROY

108 *"the first of its kind":* U.S. Air Force, *609 IWS: A Brief History, Oct 1995–Jun
1999*, https://securitycritics.org/wp-content/uploads/2006/03/hist-609.pdf.

108 *"any action to deny, exploit":* U.S. Air Force, *Cornerstones of Information Warfare*,
April 4, 1997, www.dtic.mil/cgi-bin/GetTRDoc?AD=ADA323807/.

111 *J-39 got its first taste of action:* On Operation Tango (though not J-39's role), see
Richard H. Curtiss, "As U.S. Shifts in Bosnia, NATO Gets Serious About
War Criminals," *Christian Science Monitor*, July 18, 1997; and interviews.

111 *more than thirty thousand NATO troops:* NATO, "History of the NATO-led
Stabilisation Force (SFOR) in Bosnia and Herzegovina," http://www.nato.
int/sfor/docu/d981116a.htm.

117 *"at once a great success":* Admiral James O. Ellis, "A View from the Top,"
PowerPoint presentation, n.d., http://www.slideserve.com/nili/a-view-from
-the-top-admiral-james-o-ellis-u-s-navy-commander-in-chief-u-s-naval-
forces-europe-commander-allied.

CHAPTER 8: TAILORED ACCESS

120 *In the summer of 1998:* The Air Force tried to take ownership of Joint Task
Force-Computer Network Defense, arguing that its Information Warfare
Center had unique resources and experience for the job, but Art Money and
John Hamre thought it needed to be an organization that either included all
services or transcended them. (Interviews.)

122 *So, on April 1, 2000:* U.S. Space Command, "JTF-GNO History—The Early Years of Cyber Defense," Sept. 2010; and interviews.

122 *A systematic thinker who liked:* GEDA is cited by Richard Bejtlich, "Thoughts on Military Service," *TAO Security* blog, Aug. 3, 2006, http://taosecurity. blogspot.com/2006/08/thoughts-on-military-service.html; and interviews.

123 *Suddenly, if just to stake a claim:* William M. Arkin, "A Mouse That Roars?," *Washington Post*, June 7, 1999; Andrew Marshall, "CIA Plan to Topple Milosevic 'Absurd,'" *The Independent*, July 8, 1999; and interviews.

124 *To keep NSA at the center of this universe:* NSA/CSS, *Transition 2001*, Dec. 2000, http://www2.gwu.edu/~nsarchiv/NSAEBB/NSAEBB24/nsa25.pdf; George Tenet, CIA Director, testimony, Senate Select Committee on Government Affairs, June 24, 1998, https://www.cia.gov/news-information/speeches -testimony/1998/dci_testimony_062498.html; Arkin, "A Mouse That Roars?"; and interviews.

126 *The report was written by the Technical Advisory Group:* Much of the section on TAG comes from interviews; the TAG report is mentioned in Douglas F. Garthoff, *Directors of Central Intelligence as Leaders of the U.S. Intelligence Community, 1946–2005* (Washington, D.C.: CIA Center for the Study of Intelligence, 2005), 273.

127 *The Senate committee took his report very seriously:* Senate Select Committee on Intelligence, *Authorizing Appropriations for Fiscal Year 2001 for the Intelligence Activities of the United States Government*, Senate Rept. 106-279, 106th Congress, May 4, 2000, https://www.congress.gov/congressional-report/106th-con gress/senate-report/279/1; and interviews.

128 *"poorly communicated mission":* NSA/CSS, *External Team Report: A Management Review for the Director, NSA*, Oct. 22, 1999, http://fas.org/irp/nsa/106handbk. pdf; and interviews.

129 *"is a misaligned organization":* NSA/CSS, "New Enterprise Team (NETeam) Recommendations: The Director's Work Plan for Change," Oct. 1, 1999, http://cryptome.org/nsa-reorg-net.htm.

130 *On November 15, he inaugurated:* Seymour M. Hersh, "The Intelligence Gap," *The New Yorker*, Dec. 6, 1999; and interviews.

130 *The NSA's main computer system crashed:* "US Intelligence Computer Crashes for Nearly 3 Days," CNN.com, Jan. 29, 2000, http://edition.cnn.com/2000 /US/01/29/nsa.computer/; and interviews.

132 *He called the new program Trailblazer:* NSA Press Release, "National Security Agency Awards Concept Studies for Trailblazer," April 2, 2001, https://www. nsa.gov/public_info/press_room/2001/trailblazer.shtml; Alice Lipowicz,

"Trailblazer Loses Its Way," *Washington Technology*, Sept. 10, 2005, https://washingtontechnology.com/articles/2005/09/10/trailblazer-loses-its-way.aspx.

132 *SAIC was particularly intertwined:* Siobhan Gorman, "Little-Known Contractor Has Close Ties with Staff of NSA," *Baltimore Sun*, Jan. 29, 2006, http://articles.baltimoresun.com/2006-01-29/news/0601290158_1_saic-information-technology-intelligence-experts; "Search Top Secret America's Database of Private Spooks," *Wired*, July 19, 2010, http://www.wired.com/2010/07/search-through-top-secret-americas-network-of-private-spooks/.

135 *In the coming years, TAO's ranks would swell:* "Inside TAO: Documents Reveal Top NSA Hacking Unit," *Der Spiegel*, Dec. 29, 2013, http://www.spiegel.de/international/world/the-nsa-uses-powerful-toolbox-in-effort-to-spy-on-global-networks-a-940969.html.

135 *These devices—their workings:* Matthew M. Aid, "Inside the NSA's Ultra-Secret China Hacking Group," *Foreign Policy*, June 10, 2013.

136 *One device, called LoudAuto:* The names of these programs come from a fifty-eight-page TAO catalogue of tools and techniques, among the many documents leaked by former NSA contractor Edward Snowden. No U.S. newspaper or magazine reprinted the list (the reporters and editors working the story considered it genuinely damaging to national security), but *Der Spiegel* did, in its entirety (Jacob Appelbaum, Judith Horchert, and Christian Stöcker, "Shopping for Spy Gear: Catalog Advertises NSA Toolbox," Dec. 29, 2013), and computer security analyst Bruce Schneier subsequently reprinted each item, one day at a time, on his blog.

137 *As hackers and spies discovered vulnerabilities:* "Inside TAO."

137 *In the ensuing decade, private companies:* For more on zero-day exploits, see Neal Ungerleider, "How Spies, Hackers, and the Government Bolster a Booming Software Exploit Market," *Fast Company*, May 1, 2013; Nicole Perlroth and David E. Sanger, "Nations Buying as Hackers Sell Flaws in Computer Code," *New York Times*, July 13, 2013; Kim Zetter, *Countdown to Zero Day: Stuxnet and the Launch of the World's First Digital Weapon* (New York: Crown, 2014). Specific stories come from interviews.

140 *During the first few months of Bush's term:* Richard A. Clarke, *Against All Enemies* (New York: Free Press, 2004); Steve Coll, *Ghost Wars: The Secret History of the CIA, Afghanistan, and Bin Laden, from the Soviet Invasion to September 10, 2001* (New York: Penguin, 2004), 435.

140 *On the day of the 9/11 attacks:* Robin Wright, "Top Focus Before 9/11 Wasn't on Terrorism," *Washington Post*, April 1, 2004.

141 *Rice let him draft:* Executive Order 13226—President's Council of Advisors on Science and Technology, Sept. 30, 2001, http://www.gpo.gov/fdsys/pkg/WCPD-2001-10-08/pdf/WCPD-2001-10-08-Pg1399.pdf; background, town halls, etc. come from interviews.

142 *As it turned out, the final draft:* President George W. Bush, *The National Strategy to Secure Cyberspace,* Feb. 2003, https://www.us-cert.gov/sites/default/files/publications/cyberspace_strategy.pdf.

CHAPTER 9: CYBER WARS

145 *When General John Abizaid:* For more on Abizaid and the Iraq War, see Fred Kaplan, *The Insurgents: David Petraeus and the Plot to Change the American Way of War* (New York: Simon & Schuster, 2013), esp. 182; the rest of this section comes from interviews.

148 *Meanwhile, Secretary of Defense Donald Rumsfeld:* See ibid., Ch. 4.

151 *Seventeen years had passed:* https://www.nsa.gov/about/leadership/former_directors.shtml.

151 *That same month, Rumsfeld signed:* Dana Priest and William Arkin, *Top Secret America: The Rise of the New American Security State* (New York: Little, Brown, 2011), 236.

152 *A few years earlier, when Alexander:* The section on the Alexander-Hayden feud and James Heath's experiment at Fort Belvoir comes from interviews. Some material on Heath also comes from Shane Harris, "The Cowboy of the NSA," *Foreign Policy,* Sept. 2013; and Shane Harris, *The Watchers: The Rise of America's Surveillance State* (New York: Penguin, 2010), 99, 135. Some have reported that Alexander designed the Information Dominance Center's command post to look like the captain's deck on *Star Trek,* but in fact it was set up not by Alexander or even by Noonan, but rather by Noonan's predecessor, Major General John Thomas. (Ryan Gallagher, "Inside the U.S. Army's Secretive *Star Trek* Surveillance Lair," *Slate,* Sept. 18, 2013, http://www.slate.com/blogs/future_tense/2013/09/18/surveil liance_and_spying_does_the_army_have_a_star_trek_lair.html; and interviews.)

155 *But Alexander won over Rumsfeld:* Most of this comes from interviews, but the transfer of data in June 2001 is also noted in Keith Alexander, classified testimony before House Permanent Select Committee on Intelligence, Nov. 14, 2001, reprinted in U.S. Army Intelligence and Security Command, *Annual Command History, Fiscal Year 2001,* Sept. 30, 2002 (declassified through Freedom of Information Act).

155n *Ironically, while complaining:* For details on Stellar Wind, see Barton Gellman, "U.S. Surveillance Architecture Includes Collection of Revealing Internet, Phone Metadata," *Washington Post*, June 15, 2013, and, attached on the *Post* website, the top secret draft of an inspector general's report on the program, http://apps .washingtonpost.com/g/page/world/national-security-agency-inspector -general-draft-report/277/.

156 *Trailblazer had consumed $1.2 billion:* Siobhan Gorman, "System Error," *Baltimore Sun*, Jan. 29, 2006, http://articles.baltimoresun.com/2006-01-29/ news/0601280286_1_intelligence-experts-11-intelligence-trailblazer; Alice Lipowicz, "Trailblazer Loses Its Way," *Washington Technology*, Sept. 10, 2005, http://washingtontechnology.com/articles/2005/09/10/trailblazer-loses -its-way.aspx; and interviews.

157 *Turbulence consisted of nine smaller systems:* Robert Sesek, "Unraveling NSA's Turbulence Programs," Sept. 15, 2014, https://robert.sesek.com/2014/9 /unraveling_nsa_s_turbulence_programs.html; and interviews.

158 *RTRG got under way:* This comes mainly from interviews, but also from Bob Woodward, *Obama's Wars* (New York: Simon & Schuster, 2010), 10; Ellen Nakashima and Joby Warrick, "For NSA Chief, Terrorist Threat Drives Passion to 'Collect It All,'" *Washington Post*, July 14, 2013; Shane Harris, *@War: The Rise of the Military-Internet Complex* (New York: Houghton Mifflin Harcourt, 2014), Ch. 2.

160 *In 2007 alone, these sorts of operations:* "General Keith Alexander Reveals Cybersecurity Strategies and the Need to Secure the Infrastructure," Gartner Security and Risk Management Summit, June 23–26, 2014, http://blogs .gartner.com/security-summit/announcements/general-keith-alexander -reveals-cybersecurity-strategies-and-the-need-to-secure-the-infrastructure /; and interviews.

106 *The effect was not decisive:* For more on this point, see Kaplan, *The Insurgents*, esp. Ch. 19.

160 *On September 6:* David A. Fulghum, "Why Syria's Air Defenses Failed to Detect Israelis," *Aviation Week & Space Technology*, Nov. 12, 2013; Erich Follath and Holger Stark, "The Story of 'Operation Orchard': How Israel Destroyed Syria's Al Kibar Nuclear Reactor," *Der Spiegel*, Nov. 2, 2009, http://www. spiegel.de/international/world/the-story-of-operation-orchard-how-isra-el-destroyed-syria-s-al-kibar-nuclear-reactor-a-658663.html; Richard A. Clarke and Robert A. Knake, *Cyber War* (New York: HarperCollins, 2010), 1–8; Robin Wright, "N. Koreans Taped at Syrian Reactor," *Washington Post*, April 24, 2008; "CIA Footage in Full," BBC News, April 24, 2008, http:// news.bbc.co.uk/2/hi/7366235.stm; and interviews.

161 *They did so with a computer program called Suter:* Fulghum, "Why Syria's Air Defenses Failed to Detect Israelis"; and interviews. There was some controversy over whether the target was really a nuclear reactor, but in retrospect the evidence seems indisputable. Among other things, the International Atomic Energy Agency found, in soil samples it collected around the bombed reactor, "a significant number of anthropogenic natural uranium particles (i.e., produced as a result of chemical processing)." (Follath and Stark, "The Story of 'Operation Orchard.' ")

162 *Four and a half months earlier:* "War in the Fifth Domain," *The Economist*, July 1, 2010, http://www.economist.com/node/16478792; Andreas Schmidt, "The Estonian Cyberattacks," in Jason Healey, ed., *A Fierce Domain*, 174–93; Clarke and Knake, *Cyber War*, 12–16.

164 *On August 1, 2008, Ossetian separatists:* U.S. Cyber Consequences Unit, *Overview by the US-CCU of the Cyber Campaign Against Georgia in August of 2008* (Aug. 2009), http://www.registan.net/wp-content/uploads/2009/08 /US-CCU-Georgia-Cyber-Campaign-Overview.pdf; Andreas Hagen, "The Russo-Georgian War, 2008," in Healey, ed., *A Fierce Domain*, 194–204; Government of Georgia, Ministry of Foreign Affairs, *Russian Invasion of Georgia: Russian Cyberwar on Georgia* (Nov. 10, 2008), http://www.mfa.gov.ge /files/556_10535_798405_Annex87_CyberAttacks.pdf.

166 *On March 4, 2007, the Department of Energy:* The background of the test comes from interviews. See also "Mouse Click Could Plunge City into Darkness, Experts Say," CNN, Sept. 27, 2007, http://www.cnn.com/2007/US/09/27 /power.at.risk/index.html; Kim Zetter, *Countdown to Zero Day: Stuxnet and the Launch of the World's First Digital Weapon* (New York: Crown, 2014), Ch. 9.

168 *Almost instantly, the generator shook:* For the video, see https://www.youtube. com/watch?v=fJyWngDco3g.

168 *In 2000, a disgruntled former worker:* Zetter, *Countdown to Zero Day*, 135ff.

CHAPTER 10: BUCKSHOT YANKEE

171 *When the position was created:* Fred Kaplan, "The Professional," *New York Times Magazine*, Feb. 10, 2008.

173 *So McConnell's briefing:* The date of the meeting comes from "NSC 05/16/2007-Cyber Terror" folder, NSC Meetings series, National Security Council-Records and Access Management Collection, George W. Bush Presidential Library (folder obtained through Freedom of Information Act). The substance of the meeting (which was not declassified) comes from interviews.

174 *Bush quickly got the idea*: This is based on interviews, though it's also covered
      in Shane Harris, *@War: The Rise of the Military-Internet Complex* (New York:
      Houghton Mifflin Harcourt, 2014), Ch. 2.

177 *But the task proved unwieldy:* William Jackson, "DHS Coming Up Short
      on Einstein Deployment," *GCN*, May 13, 2003, http://gcn.com/arti
      cles/2013/05/13/dhs-einstein-deployment.aspx; and interviews.

178 *On January 9, 2008:* President George W. Bush, National Security Presi-
      dential Directive (NSPD) 54, "Cyber Security Policy," Jan. 8, 2008, http://
      www.fas.org/irp/offdocs/nspd/nspd-54.pdf. The background comes from
      interviews.

179 *Meanwhile, Homeland Security upgraded Einstein:* Steven M. Bellovin et al., "Can
      It Really Work? Problems with Extending Einstein 3 to Critical Infrastruc-
      ture," *Harvard National Security Journal*, Vol. 3, Jan. 2011, http://harvardnsj
      .org/wp-content/uploads/2012/01/Vol.-3_Bellovin_Bradner_Diffie_Landau
      _Rexford.pdf; and interviews.

180 *Alexander put out the word:* Alexander cited the "Maginot Line" analogy many
      times; see for instance, "Defenses Against Hackers Are Like the 'Maginot
      Line,' NSA Chief Says," Blog, *WSJ Tech*, Jan. 13, 2012, http://blogs.wsj.com
      /digits/2012/01/13/u-s-business-defenses-against-hackers-are-like-the
      -maginot-line-nsa-chief-says/; and interviews.

181 *The pivotal moment:* The section on Buckshot Yankee comes mainly from in-
      terviews, but also from Karl Grindal, "Operation Buckshot Yankee," in Jason
      Healey, ed., *A Fierce Domain: Conflict in Cyberspace 1986 to 2012* (Washington,
      D.C.: Atlantic Council, 2013); Harris, *@War*, Ch. 9; William J. Lynn III,
      "Defending a New Domain: The Pentagon's Cyberstrategy," *Foreign Affairs*,
      Sept./Oct. 2010.

184 *When he first took the job:* For more on Gates as defense secretary, see Kaplan,
      "The Professional"; and Kaplan, *The Insurgents: David Petraeus and the Plot to
      Change the American Way of War* (New York: Simon & Schuster, 2013), Ch.
      18.

186 *On June 23, 2009:* U.S. Dept. of Defense, "U.S. Cyber Command Fact
      Sheet," May 25, 2010, http://www2.gwu.edu/~nsarchiv/NSAEBB
      /NSAEBB424/docs/Cyber-038.pdf.

186 *On July 7, 2010, Gates had lunch:* This section comes mainly from interviews,
      though the plan is briefly mentioned, along with the dates of the two meet-
      ings, in Robert Gates, *Duty: Memoirs of a Secretary at War* (New York: Alfred
      A. Knopf, 2014), 450–51.

188 *"war zone":* This section is based mainly on interviews, though in a Reuters
      profile, upon her resignation in 2013, Lute said, "The national narrative on

cyber has evolved. It's not a war zone, and we certainly cannot manage it as if it were a war zone. We're not going to manage it as if it were an intelligence program or one big law-enforcement operation." (Joseph Menn, "Exclusive: Homeland Security Deputy Director to Quit; Defended Civilian Internet Role," Reuters, April 9, 2013, http://www.reuters.com/article/2013/04/09 /us-usa-homeland-lute-idUSBRE9380DL20130409.)

188   *In the end, they approved Brown:* The watered-down version of the arrangement, "Memorandum of Agreement Between the Department of Homeland Security and the Department of Defense Regarding Cybersecurity," signed by Gates on Sept. 24 and by Napolitano on Sept. 27, 2010, can be found at http://www.defense.gov/news/d20101013moa.pdf.

CHAPTER 11: "THE WHOLE HAYSTACK"

192   *The hearings led to the passage:* The section of FISA dealing with electronic surveillance is 50 U.S.C. 1802(a).

192   *After the attacks of September 11:* A good summary is Edward C. Liu, "Amendments to the Foreign Intelligence Surveillance Act (FISA) Extended Until June 1, 2015," Congressional Research Service, June 16, 2011, https://www. fas.org/sgp/crs/intel/R40138.pdf.

193   *"badly out of date":* "The President's Radio Address," July 28, 2007, *Public Papers of the Presidents of the United States: George W. Bush, 2007, Book II* (Washington, D.C.: US Government. Printing Office, 2007), 1027–28, http://www.gpo.gov/fdsys/pkg/PPP-2007-book2/html/PPP-2007-book2-doc-pg1027.htm.

193   *"electronic surveillance of" an American:* Text of the Protect America Act of 2007, https://www.govtrack.us/congress/bills/110/s1927/text.

195   *"connect the dots":* For instance, see *The 9/11 Commission Report,* 408 and passim, http://www.9-11commission.gov/report/911Report.pdf.

197   *"the whole haystack":* The metaphor was first used by a "former intelligence officer" quoted in Ellen Nakashima and Joby Warrick, "For NSA Chief, Terrorist Threat Drives Passion to 'Collect It All,' " *Washington Post,* July 14, 2013. But Alexander was known to use the phrase, too. (Interviews.)

199   *Still, on February 9:* White House press release, Feb. 9, 2009, http://www. whitehouse.gov/the_press_office/AdvisorsToConductImmediateCyberSe curityReview/.

199   *It took longer than sixty days:* White House press release, May 29, 2009, http://www. whitehouse.gov/the-press-office/cybersecurity-event-fact-sheet-and-expect ed-attendees.

199   *It read uncannily like:* White House, *Cyberspace Policy Review*, http://www.
      whitehouse.gov/assets/documents/Cyberspace_Policy_Review_final.pdf;
      quotes come from i, iv, v, vi.
200   *"share the responsibility":* Ibid., 17.
200   *"this cyber threat":* White House, "Remarks by the President on Securing the
      Nation's Cyber Infrastructure," East Room, May 29, 2009.

CHAPTER 12: "SOMEBODY HAS CROSSED THE RUBICON"

203   *George W. Bush personally briefed:* David Sanger, *Confront and Conceal* (New
      York: Crown, 2012), xii, 190, 200–203.
203   *The operation had been set in motion:* Ibid., 191–93.
204   *In their probes:* Ibid., 196ff; Kim Zetter, *Countdown to Zero Day: Stuxnet and
      the Launch of the World's First Digital Weapon* (New York: Crown, 2014),
      Ch. 1.
205   *This would be a huge operation:* Ellen Nakashima and Joby Warrick, "Stuxnet
      Was Work of U.S. and Israeli Experts, Officials Say," *Washington Post*, June
      2, 2012.
205   *uninterruptible power supplies:* Zetter, *Countdown to Zero Day*, 200–201.
205   *A multipurpose piece of malware:* Ibid., 276–79. Much of Zetter's information
      comes from the computer virus specialists at Symantec and Kaspersky Lab
      who discovered Stuxnet. A typical malicious code took up, on average, about
      175 lines. (Interviews.)
206   *To get inside the controls:* Ibid., 90, 279.
206   *It took eight months:* Sanger, *Confront and Conceal*, 193.
206   *At the next meeting:* Ibid., xii.
206   *There was one more challenge:* Ibid., 194–96; and interviews. It has not yet been
      revealed who installed the malware-loaded thumb drives on the Iranian
      computers. Some speculate that it was an Israeli agent working at Natanz,
      some that a foreign agent (possibly with the CIA's Information Operations
      Center) infiltrated the facility, some say that contaminated thumb drives
      were spread around the area until someone unwittingly inserted one into a
      computer.
207   *Not only would the malware:* Zetter, *Countdown to Zero Day*, 61, 117, 123.
208   *Once in the White House:* Ibid., 202.
209   *but this particular worm was programmed:* Ibid., 28.
209   *Obama phoned Bush to tell him:* In his memoir, *Duty* (New York: Alfred
      A. Knopf, 2014), 303, Robert Gates writes that "about three weeks after"
      Obama's inauguration, "I called Bush 43 to tell him that we had had a

significant success in a covert program he cared about a lot." Soon after, "Obama told me he was going to call Bush and tell him about the covert success." Gates doesn't say that the classified program was Stuxnet, but it's clear from the context—and from other sections of the book where he mentions a classified program related to Iran (190–91) and denounces the leak (328)—that it is.

209 *In March, the NSA shifted its approach:* Zetter, *Countdown to Zero Day*, 303.

209 *The normal speed:* David Albright, Paul Brannan, and Christina Walrond, "ISIS Reports: Stuxnet Malware and Natanz" (Washington, D.C.: Institute for Science and International Security), Feb. 15, 2011, http://isis-online.org/uploads/isis-reports/documents/stuxnet_update_15Feb2011.pdf.

209 *They'd experienced technical problems:* An unclassified version of a 2007 National Intelligence Estimate noted that Iran was experiencing "significant technical problems operating" centrifuges ("Key Judgments from a National Intelligence Estimate on Iran's Nuclear Activity," reprinted in *New York Times*, Dec. 4, 2007); this was well before Stuxnet was activated.

209 *By the start of 2010:* Zetter, *Countdown to Zero Day*, 1–3. Similar estimates are in Albright et al., "ISIS Reports: Stuxnet Malware and Natanz."

210 *President Obama—who'd been briefed:* During briefings on Olympic Games, large foldout maps of the Natanz reactor were spread across the Situation Room (Sanger, *Confront and Conceal*, 201).

210 *Almost at once:* Michael Joseph Gross, "A Declaration of Cyber-War," *Vanity Fair*, February 28, 2011. For more details, see Nicholas Falliere, Liam O. Murchu, and Eric Chien, "Symantec Security Response: W32.Stuxnet Dossier," https://www.symantec.com/content/en/us/enterprise/media/security_response/whitepapers/w32_stuxnet_dossier.pdf; David Kushner, "The Real Story of Stuxnet," *IEEE Spectrum*, Feb. 26, 2013, http://spectrum.ieee.org/telecom/security/the-real-story-of-stuxnet; Eugene Kaspersky, "The Man Who Found Stuxnet—Sergey Ulasen in the Spotlight," *Nota Bene*, Nov. 2, 2011, http://eugene.kaspersky.com/2011/11/02/the-man-who-found-stuxnet-sergey-ulasen-in-the-spotlight/.

210 *Microsoft issued an advisory:* "Microsoft Security Bulletin MS10—046—Critical: Vulnerability in Windows Shell Could Allow Remote Execution," Aug. 2, 2010 (updated Aug. 24, 2010), https://technet.microsoft.com/en-us/library/security/ms10-046.aspx; Zetter, *Countdown to Zero Day*, 279.

210 *By August, Symantec had uncovered:* Nicolas Falliere, "Stuxnet Introduces the First Known Rootkit for Industrial Control Systems," Symantec Security Response Blog, Aug. 6, 2010, http://www.symantec.com/connect/blogs/stuxnet-introduces-first-known-rootkit-scada-devices.

210   *In September, a German security researcher:* Sanger, *Confront and Conceal*, 205–6; Joseph Gross, "A Declaration of Cyber-War."

210   *At that point, some of the American software sleuths:* Zetter, *Countdown to Zero Day*, 187–89; and interviews.

211   *When Obama learned:* Ibid., 357.

211   *The postmortem indicated:* David Sanger, "Obama Order Sped Up Wave of Cyberattacks Against Iran," *New York Times*, June 1, 2012.

211   *"offensive capabilities in cyber space":* Quoted in Richard A. Clarke and Robert K. Knake, *Cyber War* (New York: HarperCollins, 2010), 44–47.

211   *"cyber-offensive teams":* Zachary Fryer-Biggs, "U.S. Sharpens Tone on Cyber Attacks from China," *DefenseNews*, March 18, 2013, http://mobile.defense-news.com/article/303180021; and interviews.

213   *In Obama's first year as president:* Choe Sang-Hun and John Markoff, "Cyberattacks Jam Government and Commercial Web Sites in U.S. and South Korea," *New York Times*, July 18, 2009; Clarke and Knake, *Cyber War*, 23–30.

213   *A year and a half later:* Zetter, *Countdown to Zero Day*, 276–79.

213   *Four months after that:* "Nicole Perlroth, "In Cyberattack on Saudi Firm, U.S. Sees Iran Firing Back," *New York Times*, Oct. 23, 2013.

213   *"demonstrated a clear ability":* "Iran—Current Topics, Interaction with GCHQ: Director's Talking Points," April 2013, quoted and linked in Glenn Greenwald, "NSA Claims Iran Learned from Western Cyberattacks," *The Intercept*, Feb. 10, 2015, https://firstlook.org/theintercept/2015/02/10/nsa-iran-devel oping-sophisticated-cyber-attacks-learning-attacks/. The document comes from the cache leaked by Edward Snowden. The essential point is confirmed by interviews.

214   *At what point, he asked:* Gates, *Duty*, 451; and interviews.

215   *"Previous cyber-attacks had effects":* Sanger, *Confront and Conceal*, 200.

216   *"Trilateral Memorandum of Agreement":* The memorandum of agreement is mentioned in a footnote in Barack Obama, Presidential Policy Directive, PPD-20, "U.S. Cyber Operations Policy," Oct. 2012, https://www.fas.org/irp/offdocs/ppd/ppd-20.pdf. PPD-20 is among the documents leaked by Edward Snowden.

218   *An action report on the directive:* This is noted in boldfaced brackets in the copy of the document that Snowden leaked.

219   *"You can't have something that's a secret":* Andrea Shalal-Esa, "Ex-U.S. General Urges Frank Talk on Cyber Weapons," Reuters, Nov. 6, 2011, http://www.reuters.com/article/2011/11/06/us-cyber-cartwright-idUSTR-E7A514C20111106.

219 *"the authority to develop":* William B. Black Jr., "Thinking Out Loud About Cyberspace," *Cryptolog,* Spring 1997 (declassified Oct. 2012), http://cryp tome.org/2013/03/cryptolog_135.pdf. Black's precise title at the NSA was special assistant to the director for information warfare.

CHAPTER 13: SHADY RATS

221 *"rebalancing its global posture":* Thomas Donilon, speech, Asia Society, New York City, March 11, 2013, http://asiasociety.org/new-york/complete-tran script-thomas-donilon-asia-society-new-york.

222 *Then on February 18, Mandiant:* Mandiant, *APT1: Exposing One of China's Cyber Espionage Units,* Feb. 18, 2013, http://intelreport.mandiant.com/Man diant_APT1_Report.pdf.

223 *The* Times *ran a long front-page story:* David Sanger, David Barboza, and Nicole Perlroth, "Chinese Army Unit Is Seen as Tied to Hacking Against U.S.," *New York Times,* Feb. 18, 2013. The Chinese response ("irresponsible," "unprofessional," etc.) is quoted in the same article.

224 *As early as 2001:* Nathan Thornburgh, "The Invasion of the Chinese Cyberspies (And the Man Who Tried to Stop Them)," *Time,* Sept. 5, 2005; Adam Segal, "From Titan Rain to Byzantine Hades: Chinese Cyber Espionage," in Jason Healey, ed., *A Fierce Domain: Conflict in Cyberspace, 1986–2012* (Washington, D.C.: Atlantic Council/Cyber Conflict Studies Association, 2013), 165–93; and interviews.

224 *"information confrontation":* Bryan Krekel, Patton Adams, and George Bakos, *Occupying the Information High Ground,* Prepared for the U.S.-China Economic and Security Review Commission (Northrop Grumman Corporation, March 7, 2012), 9–11. http://www2.gwu.edu/~nsarchiv/NSAEBB/NSAEBB424/docs/Cyber-066.pdf

224 *By the end of the decade:* Ibid., 24–28, 40, 45–46; and interviews.

225 *he had written his doctoral dissertation:* It was published as Gregory J. Rattray, *Strategic Warfare in Cyberspace* (Cambridge: MIT Press, 2001); the rest of this section is from interviews.

225 *The typical Chinese hack started off:* Dmitri Alperovitch, McAfee White Paper, "Revealed: Operation Shady RAT," n.d., http://www.mcafee.com/us/re sources/white-papers/wp-operation-shady-rat.pdf; Ellen Nakashima, "Report on 'Operation Shady RAT' Identifies Widespread Cyber-Spying," *Washington Post,* Aug. 3, 2011; Michael Joseph Gross, "Exclusive: Operation Shady RAT—Unprecedented Cyber-espionage Campaign and Intellectual-

Property Bonanza," *Vanity Fair*, Sept. 2011; Segal, "From Titan Rain to Byzantine Hades: Chinese Cyber Espionage," 168.

228   *On June 6,* The Washington Post *and* The Guardian*:* "Verizon Forced to Hand Over Telephone Data—Full Court Ruling," *The Guardian*, June 5, 2013, accompanying Glenn Greenwald, "NSA Collecting Phone Records of Millions of Verizon Customers Daily," *The Guardian*, June 6, 2013; "NSA Slides Explain the Prism Data-Collection Program," *Washington Post*, June 6, 2013, which accompanied Barton Gellman and Laura Poitras, "U.S., British Intelligence Mining Data from Nine U.S. Internet Companies in Broad Secret Program," *Washington Post*, June 7, 2013; Glenn Greenwald and Ewen MacAskill, "NSA Prism Program Taps in to User Data of Apple, Google, and others," *The Guardian*, June 7, 2013. *The Guardian* and the *Post*, which both had Snowden documents, were locked in a fierce competition over who could publish first. The *Guardian*'s Verizon story went online June 5, then appeared in its print edition June 6. The first *Post* story went online June 6, then in print June 7. For a list of all the *Post*'s Snowden-based stories, see http://dewitt.sanford .duke.edu/gellmanarticles/.

228   *These were the first of many stories:* For the journalists' accounts of their encounters with Snowden, see "Live Chat: NSA Surveillance: Q&A with Reporter Barton Gellman," July 15, 2014, http://live.washingtonpost.com/ nsa-surveillance-bart-gellman.html; and Laura Poitras's documentary film, *CitizenFour*, 2014. For critical views of Snowden, see Fred Kaplan, "Why Snowden Won't (and Shouldn't) Get Clemency," *Slate*, Jan. 3, 2014, http:// www.slate.com/articles/news_and_politics/war_stories/2014/01/edward _snowden_doesn_t_deserve_clemency_the_nsa_leaker_hasn_t_proved _he.html; Mark Hosenball, "NSA Memo Confirms Snowden Scammed Passwords from Colleagues," Reuters, Feb. 13, 2014, http://www.reuters. com/article/2014/02/13/us-usa-security-idUSBREA1C1MR20140213; George Packer, "The Errors of Edward Snowden and Glenn Greenwald," *Prospect*, May 22, 2014, http://www.prospectmagazine.co.uk/features/the-errors-of-edward-snowden-and-glenn-greenwald.

229   *From that point on, the Chinese retort:* At a later summit, in September 2015, Obama and Xi agreed not to "conduct or knowingly support" cyber theft of "intellectual property" with the "intent of providing competitive advantage to companies or commercial sectors." The language was loose: "knowingly support" would still allow "tolerate," and an action's "intent" can be briskly denied. In any case, the U.S. doesn't conduct *this* type of cyber theft (it doesn't need Chinese trade secrets), and Xi still (absurdly) denies government in-

volvement. And the agreement doesn't cover other forms of cyber attacks or cyber espionage, not least because the U.S. engages in them, too. Still, the deal did set up a hotline and a process for investigating malicious cyber activities. It could enable deeper cooperation down the road. White House, "Fact Sheet: President Xi Jinping's State Visit to the United States," Sept. 25, 2015, https://www.whitehouse.gov/the-press-office/2015/09/25/fact-sheet -president-xi-jinpings-state-visit-united-states.

229 *One week after the failed summit:* Lana Lam and Stephen Chen, "Exclusive: Snowden Reveals More US Cyberspying Details," *South China Morning Post,* June 22, 2013, http://www.scmp.com/news/hong-kong/article/1266777 /exclusive-snowden-safe-hong-kong-more-us-cyberspying-details-re vealed?page=all.

229 *Soon came newspaper stories:* For summary, see Kaplan, "Why Snowden Won't (and Shouldn't) Get Clemency."

229 *Fort Meade's crown jewels:* Jacob Appelbaum, Judith Horchert, and Christian Stocker, "Shopping for Spy Gear: Catalog Advertises NSA Toolbox," *Der Spiegel,* Dec. 29, 2013, http://www.spiegel.de/international/world/the-nsa -uses-powerful-toolbox-in-effort-to-spy-on-global-networks-a-940969.html.

230 *Under the surveillance system described:* The potential extent of surveillance, covered by three hops, is most clearly explained in *Liberty and Security in a Changing World: Report and Recommendations of the President's Review Group on Intelligence and Communication Technologies* (White House, Dec. 12, 2013), 103, https://www.google. com/webhp?sourceid=chrome-instant&ion=1&espv=2&ie=UTF-8#q=% 22liberty%20and%20security%22%20clarke.

231 *Following this disclosure:* For instance, General Keith Alexander, testimony, House Permanent Select Committee on Intelligence, June 18, 2013, http:// icontherecord.tumblr.com/post/57812486681/hearing-of-the-house-per manent-select-committee-on.

232 *"Does the NSA collect":* Transcribed in Glenn Kessler, "James Clapper's 'Least Untruthful' Statement to the Senate," http://www.washingtonpost.com/ blogs/fact-checker/post/james-clappers-least-untruthful-statement-to-the senate/2013/06/11/e50677a8-d2d8-11e2-a73e-826d299ff459_blog.html.

232 *The day before, he'd given Clapper's office:* Senator Ron Wyden, press release, June 11, 2013, http://www.wyden.senate.gov/news/press-releases /wyden-statement-responding-to-director-clappers-statements-about-col lection-on-americans.

232 *"I thought, though, in retrospect":* Andrea Mitchell, interview with General James Clapper, NBC-TV, June 9, 2013.

235 *"besmirching the reputation"*: Steven Burke, "Cisco Senior VP: NSA Revelations Besmirched Reputation of US Companies," CRN News, Jan. 17, 2014, http://www.crn.com/news/security/240165497/cisco-senior-vp-nsa -revelations-besmirched-reputation-of-us-companies.htm?cid=rssFeed.

235 *Merkel was outraged:* Philip Oltermann, "Germany Opens Inquiry into Claims NSA Tapped Angela Merkel's Phone," *The Guardian*, June 4, 2014.

235 *There was more than a trace:* Anthony Faiola, "Germans, Still Outraged by NSA Spying, Learn Their Country May Have Helped," *Washington Post*, May 1, 2015; Reuters, "Germany Gives Huge Amount of Phone, Text Data to US: Report," http://www.nytimes.com/reuters/2015/05/12/world/europe /12reuters-germany-spying.html.

CHAPTER 14: "THE FIVE GUYS REPORT"

237 *"a high-level group":* President Obama, press conference, Aug. 9, 2013, https://www.whitehouse.gov/the-press-Noffice/2013/08/09/remarks-presi dent-press-conference.

238 *That same day:* "Administration White Paper: Bulk Collection of Telephony Metadata Under Section 215 of the USA Patriot Act," Aug. 9, 2013, http://www.publicrecordmedia.com/wp-content/uploads/2013/08 /EOP2013_pd_001.pdf; "The National Security Agency: Missions, Authorities, Oversight and Partnerships," Aug. 9, 2013, https://www.nsa.gov/pub lic_info/_files/speeches_testimonies/2013_08_09_the_nsa_story.pdf.

239 *Sunstein had written an academic paper in 2008:* Cass R. Sunstein and Adrian Vermeule, "Conspiracy Theories" (Harvard Public Law Working Paper No. 08-03; University of Chicago Public Law Working Paper No. 199), Jan. 15, 2008, http://papers.ssrn.com/sol3/papers.cfm?abstract_id=1084585.

239 *The other Chicagoan, Geoffrey Stone:* See esp. Geoffrey R. Stone, *Perilous Times: Free Speech in Wartime from the Sedition Act of 1798 to the War on Terrorism* (New York: W. W. Norton, 2006); Geoffrey Stone, *Top Secret: When Our Government Keeps Us in the Dark* (New York: Rowman & Littlefield, 2007).

239 *Peter Swire:* peterswire.net; and interviews.

240 *"To the loved ones":* Transcript, Richard A. Clarke, testimony, 9/11 Commission, March 24, 2004, http://www.cnn.com/TRANSCRIPTS/0403/28/le.00 .html.

240 *a segment on CBS TV's* 60 Minutes: "The CBS 60 Minutes Richard Clarke Interview," http://able2know.org/topic/20967-1.

241 *Published in April 2010:* For examples of criticism, see Ryan Singel, "Richard Clarke's *Cyber War*: File Under Fiction," *Wired*, April 22, 2010.

241 *"Cyber-war, cyber-this":* Jeff Stein, "Book Review: 'Cyber War' by Richard Clarke," *Washington Post*, May 23, 2010.

242 *On August 27:* http://www.dni.gov/index.php/intelligence-community/re view-group; the substance of the meeting comes from interviews.

243 *The next morning:* The date of the first meeting at Fort Meade comes from a highly entertaining video of Geoffrey Stone delivering the "Journeys" lecture at the University of Chicago, sometime in 2014, http://chicagohuman-ities.org/events/2014/journeys/geoffrey-stone-on-the-nsa; substance of the session comes from that video and interviews.

243 *In* Cyber War, *he'd criticized:* Richard A. Clarke and Robert K. Knake, *Cyber War* (New York: HarperCollins, 2010), passim, esp. 44ff.

244 *Stone was no admirer of Snowden:* "Is Edward Snowden a Hero? A Debate with Journalist Chris Hedges and Law Scholar Geoffrey Stone," *Democracy Now*, June 12, 2013, http://www.democracynow.org/2013/6/12/is_edward _snowden_a_hero_a.; and interviews.

245 *Moreover, if the metadata revealed:* The figure of twenty-two NSA officials comes from the White House, *Liberty and Security in a Changing World: Report and Recommendations of the President's Review Group on Intelligence and Communication Technologies*, Dec. 12, 2013 (hereinafter cited as "President's Review Group"), 98, https://www.nsa.gov/civil_liberties/_files/liberty_security_prgfinalreport. pdf; the rest of this section, unless otherwise noted, comes from interviews.

245 *second hop:* A clear discussion of hops can be found in ibid., 102–3.

246 *For all of 2012:* The numbers—288, 12, and 0—are cited in ibid., 104.

246 *"Uh, hello?":* Geoffrey Stone, interview, NBC News, "Information Clearing House," Dec. 20, 2013, http://www.informationclearinghouse.info /article37174.htm; and interviews.

247 *It concerned the program known as PRISM:* This was the first news leak from Snowden, who had not yet come out as the source. See Barton Gellman and Laura Poitras, "U.S., British Intelligence Mining Data from Nine U.S. Internet Companies in Broad Secret Program," *Washington Post*, June 7, 2013; the discussion at Fort Meade comes from interviews.

247 *"the most significant tool":* Quoted in Jack Bouboushian, "Feds Ponder Risk in Preserving Spying Data," Courthouse News Service, June 6, 2014, http:// www.courthousenews.com/2014/06/06/68528.htm. The same language was later used in the NSA's Aug. 9 release on its missions and authorities (see above), as well as in a joint statement on Aug. 22, 2013 by the NSA and the Office of the Director of National Intelligence, http://www.dni.gov/index.php /newsroom/press-releases/191-press-releases-2013/917-joint-statement-nsa-and-office-of-the-director-of-national-intelligence.

247 *General Alexander had publicly claimed:* NBC News, June 27, 2013, http://usnews.nbcnews.com/_news/2013/06/27/19175466-nsa-chief-says-surveillance-programs-helped-foil-54-plots; and interviews.

249 *"selectors". . . "foreignness" . . . 52 percent:* This was also cited in Gellman and Poitras, "U.S., British Intelligence Mining Data from Nine U.S. Internet Companies in Broad Secret Program."

249 *Each year the agency's director:* President's Review Group, 138.

250 *"tens of thousands of wholly domestic communications":* Cited in ibid., 141–42.

251 *But to some of the panelists:* This comes from interviews, but the thought is expressed throughout the report, for instance, 61, 76, 113–16, 125.

252 *Morell and the staff . . . concluded:* Ibid., 144–45.

252 *However, in* none *of those fifty-three files:* Ibid., 104; and interviews.

253 *Alexander also revealed:* Ibid., 97; and interviews.

252 *"This is bullshit":* Stone, "Journeys" lecture, University of Chicago; and interviews.

256 *"reduce the risk":* President's Review Group, 118. For the other recommendations cited, see 34, 36, 86, 89.

257 *"subvert, undermine, weaken":* Ibid., 36–37.

257 *Finally, lest anyone interpret the report:* These were Recommendations Nos. 37 through 46. Ibid., 39–42.

258 *On December 13:* White House press spokesman Jay Carney cited the date in his Dec. 16 briefing, https://www.whitehouse.gov/the-press-office/2013/12/16/daily-briefing-press-secretary-12162013.

258 *"to promote public trust":* President's Review Group, 49.

259 *"Although recent disclosures":* Ibid., 75–76.

259 *"no evidence of illegality":* Ibid, 76.

259 *"the lurking danger":* Ibid., 113.

259 *"We cannot discount":* Ibid., 114.

259 *On December 18:* White House, President's Schedule, https://www.whitehouse.gov/schedule/president/2013-12-18.

259 *"We cannot prevent terrorist attacks":* "Remarks by the President on Review of Signals Intelligence," Jan. 17, 2014, https://www.whitehouse.gov/the-press-office/2014/01/17/remarks-president-review-signals-intelligence.

260 *"in the sense that there's no clear line":* Liz Gannes, "How Cyber Security Is Like Basketball, According to Barack Obama," *re/code*, Feb. 14, 2015, http://recode.net/2015/02/14/how-cyber-security-is-like-basketball-according-to-barack-obama/.

261 *The questions to be asked:* Michael Daniel, White House cybersecurity chief, re-

vealed this decision, and outlined these criteria, in his blog of April 28, 2014, headlined "Heartbleed: Understanding When We Disclose Cyber Vulnerabilities," https://www.whitehouse.gov/blog/2014/04/28/heartbleed-under standing-when-we-disclose-cyber-vulnerabilities.

262 *"unprecedented and unwarranted"*: The ruling came in the case of *ACLU v. Clapper*, http://pdfserver.amlaw.com/nlj/NSA_ca2_20150507.pdf. A lower court had ruled in favor of Clapper and thus upheld the FISA Court's concept of "relevance" and the legality of NSA bulk collection; the U.S. Court of Appeals for the 2nd Circuit in New York overturned that ruling. I analyzed the ruling and its implications in Fred Kaplan, "Mend It, Don't End It," *Slate*, May 8, 2015, http://www.slate.com/articles/news_and_poli tics/war_stories/2015/05/congress_should_revise_the_patriot_act_s_section _215_the_national_security.html.

264 *"To be clear"*: Stone published a shortened version of his talk, on the same day, as Geoffrey R. Stone, "What I Told the NSA," *Huffington Post*, March 31, 2014, http://www.huffingtonpost.com/geoffrey-r-stone/what-i-told -the-nsa_b_5065447.html; this account of his speech is based on that article and on interviews.

CHAPTER 15: "WE ARE WANDERING IN DARK TERRITORY"

265 *In the wee hours*: Most of the material on the Vegas hack is from Ben Elgin and Michael Riley, "Now at the Sands Casino: An Iranian Hack in Every Server," *Bloomberg Businessweek*, Dec. 11, 2014, http://www.bloomberg.com /bw/articles/2014-12-11/iranian-hackers-hit-sheldon-adelsons-sands-casi no-in-las-vegas; a bit is from interviews.

268 *"Guardians of Peace"*: James Cook, "Sony Hackers Have Over 100 Terabytes of Documents," *Business Insider*, Dec. 16, 2014; Mark Seal, "An Exclusive Look at Sony's Hacking Saga," *Vanity Fair*, Feb. 2015; Kevin Mandia, quoted in "The Attack on Sony," *60 Minutes*, CBS TV, Apr. 12, 2015, http://www.cbsnews.com/news/north-korean-cyberattack-on-sony-60-minutes/.

268 *Sony had been hacked before*: Keith Stuart and Charles Arthur, "PlayStation Network Hack," *The Guardian*, April 27, 2011; Jason Schreier, "Sony Hacked Again: 25 Million Entertainment Users' Info at Risk," Wired.com, May 2, 2011, http://www.wired.com/2011/05/sony-online-entertainment-hack/.

268 *The cost, in business lost*: Jason Schreier, "Sony Estimates $171 Million Loss

from PSN Hack," Wired.com, May 23, 2011, http://www.wired.com/2011/05 /sony-psn-hack-losses/.

268   *So the lessons learned in one realm:* John Gaudiosi, "Why Sony Didn't Learn from Its 2011 Hack," Fortune.com, Dec. 24, 2014, http://fortune.com/2014/12/24 /why-sony-didnt-learn-from-its-2011-hack/.

269   *"DarkSeoul":* Brandon Bailey and Youkyung Lee, "Experts Cite Similarities Between Sony Hack and 2013 South Korean Hacks," Associated Press, Dec. 4, 2014, http://globalnews.ca/news/1707716/experts-cite-similarities-between-sony-hack-and-2013-south-korean-hacks/.

269   *"mercilessly destroy":* David Tweed, "North Korea to 'Mercilessly' Destroy Makers of Rogen Film," *BloombergBusiness,* June 26, 2014, http://www .bloomberg.com/news/articles/2014-06-26/north-korea-to-mercilessly-destroy-makers-of-seth-rogan-film.

269   *In public, officials said:* "The Attack on Sony," *60 Minutes*; "NSA Chief Says Sony Attack Traced to North Korea After Software Analysis," Reuters, Feb. 19, 2015, http://www.nytimes.com/reuters/2015/02/19/technology/19reuters -nsa-northkorea-sony.html?_r=0.

269   *But the real reason:* David E. Sanger and Martin Fackler, "NSA Breached North Korean Network Before Sony Attack, Officials Say," *New York Times,* Jan. 18, 2015; and interviews.

270   *"made a mistake":* "Remarks by the President in Year-End Press Conference," White House, Dec. 19, 2014, https://www.whitehouse.gov/the-press -office/2014/12/19/remarks-president-year-end-press-conference.

270   *"not just an attack":* Statement by Secretary Johnson on Cyber Attack on Sony Pictures Entertainment, Department of Homeland Security, Dec. 19, 2014, http://www.dhs.gov/news/2014/12/19/statement-secretary-johnson-cyber-attack-sony-pictures-entertainment.

271   *On December 22:* Nicole Perlroth and David E. Sanger, "North Korea Loses Its Link to the Internet," *New York Times,* Dec. 22, 2014. That the U.S. government did not launch the attack comes from interviews.

272   *"the first aspect of our response":* Statement by the Press Secretary on the Executive Order "Imposing Additional Sanctions with Respect to North Korea," White House, Jan. 2, 2015, https://www.whitehouse.gov/the-press-of-fice/2015/01/02/statement-press-secretary-executive-order-entitled-impos-ing-additional-sanctions-respect-north-korea. The backstory on the pointed wording comes from interviews.

273   *Those who heard Gates's pitch:* In President Obama's PPD-20, "U.S. Cyber Operations Policy," one of the directives, apparently inspired by Gates's idea, reads as follows: "In coordination with the Secretaries of Defense and

Homeland Security, the AG, the DNI, and others as appropriate, shall continue to lead efforts to establish an international consensus around norms of
behavior in cyberspace to reduce the likelihood of and deter actions by other
nations that would require the United States Government to resort to" cyber
offensive operations. In a follow-on memo, summarizing actions that the
designated departments had taken so far, the addendum to this one reads:
"Action: [Department of] State; ongoing"—signifying, in other words, no
progress (http://fas.org/irp/offdocs/ppd/ppd-20.pdf).

273 *In 2014, there were almost:* The precise numbers for 2014 were 79,790 breaches,
with 2,122 confirmed data losses; for 2013, 63,437 breaches, with 1,367 losses.
Espionage was the motive for 18 percent of the breaches; of those, 27.4 percent
were directed at manufacturers, 20.2 percent at government agencies. Verizon, *2014 Data Breach Investigations Report*, April 2015, esp. introduction, 32,
52, file:///Users/fred/Downloads/rp_Verizon-DBIR-2014_en_xg%20(3).pdf.
For 2013 data: Verizon, *2013 Data Breach Investigations Report*, April 2014, file:///
Users/fred/Downloads/rp_data-breach-investigations-report-2013_en_xg.pdf.

273 *On average, the hackers stayed inside: Cybersecurity: The Evolving Nature of Cyber
Threats Facing the Private Sector,* Before the Subcommittee on Information Technology, 114th Cong. (2015). (Statement of Richard Bejtlich, FireEye Inc.) http://
oversight.house.gov/wp-content/uploads/2015/03/3-18-2015-IT-Hearing-
on-Cybersecurity-Bejtlich-FireEye.pdf.

273n *In 2013, two security researchers*: Andy Greenberg, "Hackers Remotely Kill a
Jeep on the Highway—With Me in It," *Wired*, July 21, 2015, http://www.
wired.com/2015/07/hackers-remotely-kill-jeep-highway/. A team of university researchers spelled out this vulnerability still earlier, in Stephen Check-
oway, et al., "Comprehensive Experimental Analyses of Automotive Attack
Surfaces," http://www.autosec.org/pubs/cars-usenixsec2011.pdf. The 2013
experiment by Charlie Miller and his colleague, Chris Velasek, was designed
to test that paper's proposition.

274 *"Nothing in this order":* President Barack Obama, Executive Order—Improving Critical Infrastructure Cybersecurity, Feb. 12, 2013, https://www
.whitehouse.gov/the-press-office/2013/02/12/executive-order-improving
-critical-infrastructure-cybersecurity.

275 *"disrupting or completely beating":* Department of Defense, Defense Science
Board, Task Force Report, *Resilient Military Systems and the Advanced Cyber
Threat*, Jan. 13, 2013, cover memo and executive summary, 1, http://www
.acq.osd.mil/dsb/reports/ResilientMilitarySystems.CyberThreat.pdf.

275 *Some of the task force members:* Ibid., Appendix 2; "time machine" comes from
interviews.

276   *"The network connectivity"*: Ibid., Executive Summary, 15.

276   *"built on inherently insecure architectures"*: Ibid., cover memo, 1, 31.

277   *"With present capabilities"*: Ibid.

277   *"Thus far the chief purpose"*: Bernard Brodie, *The Absolute Weapon* (New York: Harcourt Brace, 1946), 73–74, 76. For more on Brodie, and the subject generally, see Fred Kaplan, *The Wizards of Armageddon* (New York: Simon & Schuster, 1983).

278   *"Define and develop enduring"*: Barack Obama, White House, "The Comprehensive National Cybersecurity Initiative," https://www.whitehouse.gov /issues/foreign-policy/cybersecurity/national-initiative.

278   *"It took decades"*: Department of Defense, Defense Science Board, Task Force Report, *Resilient Military Systems and the Advanced Cyber Threat*, 51. Actually, in the mid-1990s, the RAND Corporation did conduct a series of war games that simulated threats and responses in cyber warfare; several included upper-midlevel Pentagon officials and White House aides as players, but no insiders took them seriously; the games came just a little bit too early to have impact. The games were summarized in Roger C. Molander, Andrew S. Riddile, Peter A. Wilson, *Strategic Information Warfare: A New Face of War* (Washington, D.C.: RAND Corporation, 1996). The dearth of impact comes from interviews.The presented a ninety-page paper, explaining how they did the hack (and spelling out disturbing implications), at the August 2015 Black Hat conference in Las Vegas (Remote Exploitation of an Unaltered Passenger Vehicle," illmatics.com//remote7. 20Car7.20Hacking.pdf).

279   *"to consider the requirements"*: Undersecretary of Defense (Acquisition, Technology, and Logistics), Memorandum for Chairman, Defense Science Board, "Terms of Reference—Defense Science Board Task Force on Cyber Deterrence," Oct. 9, 2014, http://www.acq.osd.mil/dsb/tors/TOR-2014-10-09-Cyber_Deterrence.pdf. The date of the first session and the names of the task force members come from interviews.

280   *In 2011, when Robert Gates realized*: The directive is summarized, though obliquely, in Department of Defense, *Department of Defense Strategy for Operating in Cyberspace*, July 2011, http://www.defense.gov/news/d20110714cy ber.pdf; see also Aliya Sternstein, "Military Cyber Strike Teams Will Soon Guard Private Networks," *NextGov.com*, March 21, 2013, http://www.next gov.com/cybersecurity/cybersecurity-report/2013/03/military-cyber-strike-teams-will-soon-guard-private-networks/62010/; and interviews.

282   *"biggest focus"*: Quoted in Cheryl Pellerin, "Rogers: Cybercom Defending

Networks, Nation," *DoD News*, Aug. 18, 2014, http://www.defense.gov/news/newsarticle.aspx?id=122949.

283 *"with other government agencies":* Department of Defense, *The Department of Defense Cyber Strategy*, April 2015; quotes on 5, 14, emphasis added; see also 6, http://www.defense.gov/home/features/2015/0415_cyber-strategy/Final_2015_DoD_CYBER_STRATEGY_for_web.pdf. The document clarified that the government would be responsible for deterring and possibly responding only to cyber attacks "of significant consequence," which, it added, "may include loss of life, significant damage to property, serious adverse U.S. foreign policy consequences, or serious economic impact on the United States." The terms "significant" and "serious" remained undefined—Robert Gates's question, nine years earlier, of what kind of cyber attack constitutes an act of war remained unanswered—but the finesse reflected an understanding that all such questions are ultimately political, to be decided by political leaders. It also reflected the inescapable fact that this was not just dark but untrod territory.

283 *"How do we increase":* Ellen Nakashima, "Cyber Chief: Efforts to Deter Attacks Against the US Are Not Working," *Washington Post*, March 19, 2015.

283 *"probably one or two":* Patricia Zengerle, "NSA Chief Warns Chinese Cyber Attack Could Shut U.S. Infrastructure," Reuters, Nov. 21, 2014, http://www.reuters.com/article/2014/11/21/usa-security-nsa-idUSL2N0T-B0IX20141121.

285 *"The American public":* *Liberty and Security in a Changing World:* President's Review Group, 62.

ACKNOWLEDGMENTS

I CAME UP with the idea for this book—the contract was drawn up, the research was begun, the first interviews with sources were conducted—before the world had heard of Edward Snowden; before *metadata*, *PRISM*, and *encryption* entered the banter of common conversation; before cyber attacks—launched by China, Russia, North Korea, Iran, organized crime groups and, yes, the United States government—became the stuff of headline news seemingly every day. My proposal was to write a *history* of what has broadly come to be called "cyber war," and my interest in the idea grew as the stories piled up about Snowden and the thousands of documents he leaked, because it was clear that few people, even among those who studied the documents closely (I suspect, even among those who wrote about the documents, even Snowden himself) knew that there *was* a history or, if they did, that this history stretched back not a few years but five decades, to the beginnings of the Internet itself.

This book can be seen as the third in a series of books that I've written about the interplay of politics, ideas, and personalities in modern war. The first, *The Wizards of Armageddon* (1983), was about the think-tank intellectuals who invented nuclear strategy and wove its tenets into official policy. The second, *The Insurgents* (2013), was about the intellectual Army officers who revived counterinsurgency doctrine and tried to apply it to the wars in Iraq and Afghanistan. Now, *Dark Territory* traces the players, ideas, and technology of the looming cyber wars.

On all three books, I've had the great fortune of working with Alice Mayhew, the legendary editor at Simon & Schuster, and it's to her that I owe their existence. The seeds of this book were planted during a conversation in her office either in December 2012 or January 2013 (just before or just after publication of *The Insurgents*), when, trying to nudge me into writing another book, Alice asked what the next big topic in military matters was likely to be. I vaguely replied that this "cyber" business might get serious. She asked me more questions; I answered them as fully as I could (I didn't really know a lot about the subject at the time). By the time the meeting ended, I was committed to looking into a book about cyber war— first, to see if there was a story there, a story with characters and a narrative pulse. It turned out, there was.

I thank Alice for prodding me in this direction and for asking other pointed questions at every step along the way. I thank the entire S&S team on the project: Stuart Roberts, Jackie Seow, Jonathan Evans, Maureen Cole, Larry Hughes, Ellen Sasahara, Devan Norman, and, especially, the publisher, Jonathan Karp. I thank Fred Chase for scrupulous copyediting. I thank Alex Carp and Julie Tate for diligent fact-checking (though I bear total responsibility for any errors that remain).

Additional support came from the Council on Foreign Relations, where I was the Edward R. Murrow press fellow during the year when I did much of the book's research. I thank, in particular, the fellowship's leaders, Janine Hill, Victoria Alekhine, and, my energetic assistant during the year, Aliya Medetbekova, as well as the Council's many fellows, staff specialists, and visiting speakers with whom I had spirited conversations. (I should stress that neither the Council nor anyone at the Council had any role whatsoever in the book itself, beyond providing me a nice office, stipend, and administrative assistance.)

In the course of my research, I interviewed more than one hundred people who played a role in this story, many of them several times, with follow-ups in email and phone calls. They ranged from cabinet secretaries, generals, and admirals (including six directors of the National Security Agency) to technical specialists in the hidden corridors of the security bureaucracy (not just the NSA), as well as officers, officials, aides, and analysts at every echelon in between. All of these interviews were conducted in confidence; most of the sources agreed to talk with me only under those conditions, though I should note that almost all of the book's facts (and, when it comes to historically new disclosures, *all* the facts) come from at least two sources in positions to know. I thank all of these people: this book would not exist without you.

I also thank Michael Warner, the official historian of U.S. Cyber Command, and Jason Healey and Karl Grindal of the Cyber Conflict Studies Association, whose symposiums and collections of declassified documents were instrumental in persuading me, at an early phase of the project, that there *was* a story, a history, to be told here.

This is my fifth book in thirty-three years, and they've all been guided into daylight by Rafe Sagalyn, my literary agent, who has stood by throughout as taskmaster, counselor, and friend. I thank him once again, as well as his patient assistants, Brandon Coward and Jake DeBache.

Finally, I am grateful to my friends and family for their encouragement in so many ways. I especially thank my mother Ruth Kaplan Pollack, who has always been there with support of various kinds; my wife, Brooke Gladstone, who has loomed as my best friend, life's love, and moral compass since we were both barely out of our teens; and our daughters, Sophie and Maxine, whose integrity and passion continue to astonish me.

INDEX

Abizaid, John, 159, 173
  cyber warfare as priority of, 145–48,
    149–50
Abkhazia, 164–65
Abraham Lincoln, USS, 147
Absolute Weapon, The (Brodie), 277
Adelson, Sheldon, 265–66, 267
Afghanistan, 147, 182, 199, 229
  U.S. war in, 208
Against All Enemies (Clarke), 240
agent.btz (computer worm), 182
Air Combat Command, Information
  Warfare Branch of, 107, 110
Air Force, U.S., 64, 70, 79
  Office of Special Investigations of,
    85, 87
Air Force Cryptology Support Center, 62
Air Force Information Operations
  Center, 225
Air Force Information Warfare Center
  (Air Intelligence Agency), 7, 32,
    58–59, 85, 107, 108, 110, 111, 112,
    122–23, 126, 137, 161, 176, 212, 223,
    225, 292n–93n, 296n
  Computer Emergency Response
    Team of, 62–63, 69, 73
  demon-dialing counter-C2 plan of,
    59, 64
  Network Security Monitoring
    program of, 60–61, 62
Air Force Scientific Advisory Board, 51
Air Force Systems Command, 50
Alexander, Keith, 148–49, 173, 213
  as Army deputy chief of staff for
    intelligence, 149–50
  as Army Intelligence and Security
    commander, 148, 152, 154–55, 196
  BuckShot Yankee and, 182–84
  CNE and, 180
  as CyberCom head, 189, 211
  cyber warfare expertise of, 149, 157–58

Hayden's conflicts with, 152–53, 154–55
infrastructure security and, 280–82
metadata and, 230, 231, 233, 253, 256
as NSA director, 152, 155–56, 174,
  178–81, 182–84, 185–86, 187, 189,
  204, 211, 214, 231, 244, 247, 252,
  253, 256
Obama information-sharing bill
  opposed by, 281–82
PRISM and, 247
and Snowden leaks, 231
Stuxnet attack and, 204–5, 206
Turbulence and, 157–58
al Qaeda, 140, 142–43, 147, 151, 171,
  192, 197, 240–41, 245
  U.S. drone strikes on, 201, 208
Amazon, 102
American Civil Liberties Union
  (ACLU), 239
American Electrical Power Company, 167
Anderson, Jack, 288n–89n
Andrews, Duane, 54
Andrews Air Force Base, cyber attack on,
  73, 74
AntiOnline, 77
AOL, PRISM and, 247
Apple, PRISM and, 247
Aristide, Jean-Bertrand, 58, 59
Army, U.S., 70, 79, 151
  Intelligence and Security Command
    of, 148, 152–55, 196
  Land Information Warfare Activity
    of, 32, 123
Army Intelligence Center, 148–49
ARPANET, 7–9, 276
Arquilla, John, *291n*
ASD(C3I), 6, 20, 54, 119, 125
Asia Society, Donilon's speech at,
  221–22, 226–27
Assad, Bashar al-, 160–61, 198
Assante, Michael, 166–67

323

ABOUT THE AUTHOR

Fred Kaplan writes the "War Stories" column for *Slate*. A former Pulitzer Prize-winning reporter for the *Boston Globe*, he is the author of four previous books, *The Insurgents: David Petraeus and the Plot to Change the American Way of War* (which was a Pulitzer Prize finalist), *1959: The Year Everything Changed*, *Daydream Believers: How a Few Grand Ideas Wrecked American Power*, and *The Wizards of Armageddon* (which won the Washington Monthly Political Book of the Year Award). He has a PhD in political science from MIT. He lives in Brooklyn with his wife, Brooke Gladstone.